InstantRevision

Published by HarperCollins*Publishers* Ltd
77–85 Fulham Palace Road
London W6 8JB

First published 2002
This new format edition published 2004

ISBN 0 00 717267 2

British Library Cataloguing in Publication Data
A catalogue record for this book is available from the British Library.

Edited by Eva Fairnell
Production by Katie Butler
Design by Gecko Ltd
Illustrations by Gecko Ltd
Cover design by Susi Martin-Taylor
Printed and bound by Printing Express Ltd, Hong Kong

Every effort has been made to contact the holders of copyright material, but if any have been inadvertently overlooked, the Publishers will be pleased to make the necessary arrangements at the first opportunity.

Get the most out of your Instant Revision pocket book

1. **Maximize your revision time.** You can carry this book around with you anywhere. This means you can spend any spare moments revising.
2. **Learn and remember what you need to know.** The book contains all the really important facts you need to know for your exam. All the information is set out clearly and concisely, making it easy for you to revise.
3. **Find out what you don't know.** The *Check yourself* questions help you to discover quickly and easily the topics you're good at and those you're not so good at.

What's in this book

1 The content you need for *your* AS exam

- This book covers the key content of all the specifications of the four awarding bodies: AQA, Edexcel, OCR and WJEC. The content requirements differ between the specifications.
- **Make sure you know which specification you are entered for.** Get a copy of your specification, and mark the relevant parts of each chapter in this book so that you know which parts you need.

2 Chemical calculations

- An extensive chapter on calculation questions is included, with plenty of opportunity to practise AS-type problems.
- Full explanations of all answers are given to help build your confidence in this area.

3 Definitions and equations you must know

- Many of the questions in AS exams expect you to give precise definitions and correctly balanced chemical equations. These are clearly set out for you in this book. For every organic reaction, the necessary reagents and conditions are systematically included.

4 questions – find out how much you know and improve your grade

- The *Check yourself* questions appear at the end of each short topic chapter.
- The questions are quick to answer. They are not actual exam questions but the author, who is a senior examiner, has written them in such a way that they will highlight any vital gaps in your knowledge and understanding.
- The answers are given at the back. When you have answered the questions, check your answers with those given. The author gives help with arriving at the correct answer, so if your answer is incorrect, you will know where you went wrong.
- There are marks for each question. If you score very low marks for a particular *Check yourself* page, this shows that you are weak on that topic and need to put in more revision time there.

Revise actively!

- Concentrated, active revision is much more successful than spending long periods reading through notes with half your mind elsewhere.
- The chapters in this book are quite short. For each of your revision sessions, choose a couple of topics and concentrate on reading and thinking them through for **20–30 minutes**. Then do the *Check yourself* questions. If you get a number of questions wrong, you will need to return to the topics at a later date. Some Chemistry topics are hard to grasp but, by coming back to them several times, your understanding will improve and you will become more confident about using them in the exam.
- You can use this book to revise effectively on your own – or with a friend!

The model of an atom

In a simple model of the atom, **protons** and **neutrons** are found in the **nucleus** of the atom. **Electrons** are arranged in shells around the nucleus.

Example $^{12}_{6}C$

For any individual atom, **X**, two numbers are used to describe it, $^{N}_{Z}X$. These are:

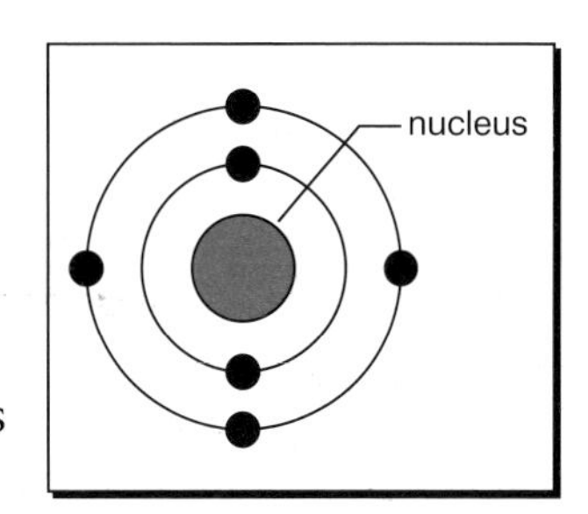

N, the mass number = number of protons + number of neutrons

and

Z, the atomic number = number of protons (in a **neutral atom**, the number of protons equals the number of **electrons**).

Example $^{12}_{6}C$

Number of protons = 6 = number of electrons

Number of neutrons = (mass number – atomic number)

Therefore number of neutrons = (12 – 6) = 6

The relative masses and charges of the three subatomic particles are:

Particle	Relative mass	Relative charge
Proton	1	1+
Neutron	1	0
Electron	$\frac{1}{1840} \approx 0$	1–

Isotopes

Isotopes are atoms of the same element that differ only in the number of neutrons.

Example $^{35}_{17}Cl$ and $^{37}_{17}Cl$

protons = 17	protons = 17
electrons = 17	electrons = 17
neutrons = 18	neutrons = 20

Therefore, isotopes have the **same** atomic number but **different** mass number.

Relative isotopic, atomic and molecular masses are measured on a scale in which the mass of an atom of carbon-12 is exactly 12 **atomic mass units (a.m.u.)**.

$$\textbf{Relative isotopic mass} = \frac{\text{the mass of one atom of a specific isotope}}{\frac{1}{12}\text{ the mass of one atom of }^{12}\text{C}}$$

Example For an atom of the isotope $^{14}_{6}C$

$$\text{Relative isotopic mass} = \frac{14}{\frac{1}{12}\times 12}\ \frac{\text{a.m.u.}}{\text{a.m.u.}}$$

$$= \frac{14}{1}$$

$$= 14\ \textbf{(no units)}$$

Relative atomic mass (A_r)

$$= \frac{\text{the average mass of one atom of an element}}{\frac{1}{12}\text{ the mass of one atom of }^{12}\text{C}}$$

The relative atomic mass is the 'weighted average' of the mass numbers of all the isotopes of a particular element.

Example Naturally occurring chlorine is 75% ^{35}Cl and 25% ^{37}Cl

The average mass of an atom of chlorine is:

$$\left(\frac{75.0}{100} \times 35\right) + \left(\frac{25.0}{100} \times 37\right) = 35.5 \text{ a.m.u.}$$

The relative atomic mass (A_r) of chlorine is therefore:

$$\frac{35.5 \quad \text{a.m.u.}}{\frac{1}{12} \times 12 \quad \text{a.m.u.}}$$

$$= \frac{35.5}{1}$$

$= 35.5$ **(no units)**

For a molecule of a substance:

Relative molecular mass (M_r)

$$= \frac{\text{the average mass of one molecule}}{\frac{1}{12} \text{ the mass of one atom of } ^{12}C}$$

The mass spectrometer

A **low-resolution mass spectrometer** may be used to work out relative atomic mass.

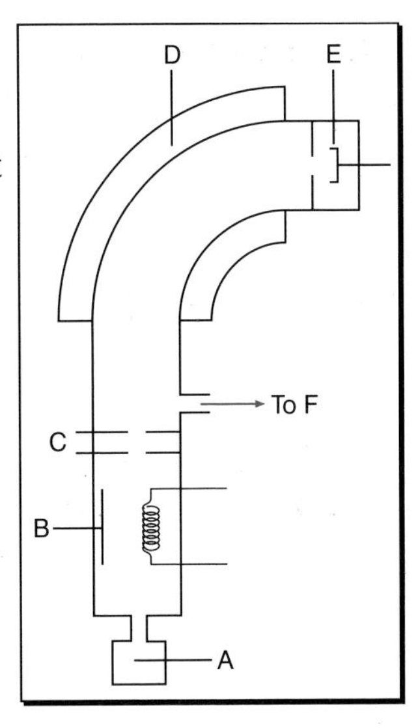

A: a vaporised sample of atoms of the element

B: positive ions are formed by electron bombardment

C: the positive ions are accelerated by an electric field

D: the positive ions are deflected by a magnetic field

E: the positive ions are detected

F: a vacuum pump, to ensure that no air molecules are present in the mass spectrometer

From the mass spectrum obtained, the relative atomic mass of an element may be calculated.

Example Neon

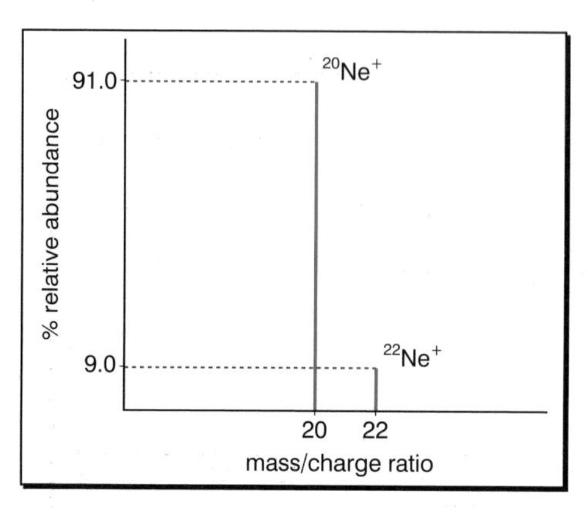

$$\text{Relative atomic mass of neon} = \left(\frac{91.0}{100} \times 20\right) + \left(\frac{9.00}{100} \times 22\right)$$
$$= 18.2 + 1.98$$
$$= 20.18$$
$$= 20.2 \text{ (3 sig. fig.)}$$

Ionisation energies

The **first ionisation energy (IE_1)** is the energy required to remove one mole of electrons from one mole of gaseous atoms of an element, to form one mole of singly charged positive ions:

$X_{(g)} \longrightarrow X^+_{(g)} + e^-$

The second ionisation energy (IE_2) is the energy required to remove one mole of electrons from one mole of singly-charged positive ions in the gas phase, to form one mole of doubly-charged positive ions:

$X^+_{(g)} \longrightarrow X^{2+}_{(g)} + e^-$

Ionisation energies are normally measured in **kJ mol^{-1}**.

Successive ionisation energies of an element provide evidence for the existence of **quantum shells**.

Example A plot of $\log_{10}$ IE against number of electrons removed, for sodium

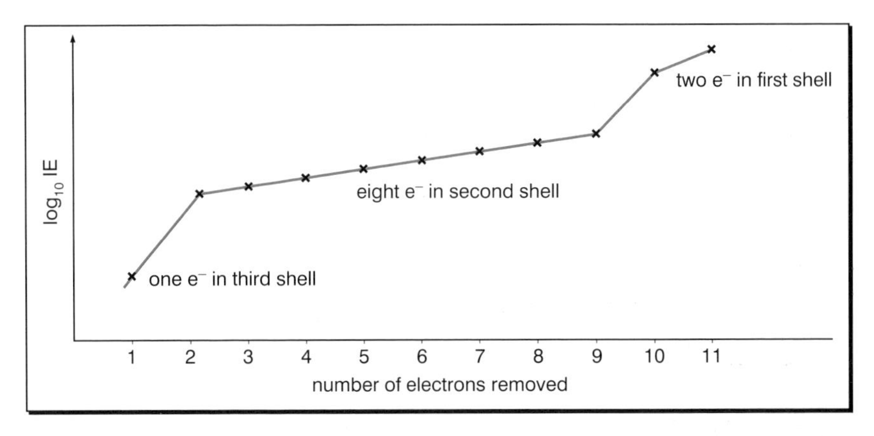

The graph provides evidence that the electronic configuration of the sodium atom is **2, 8, 1**.

Ionisation energies are influenced by:

- nuclear charge (number of protons in the nucleus);
- electron shielding (number of inner shells of electrons);
- distance of outermost electron from the nucleus (shell number of outermost electron).

From the graph above, note that:

- there are big 'jumps' between $\mathbf{IE_1} \longrightarrow \mathbf{IE_2}$ and $\mathbf{IE_9} \longrightarrow \mathbf{IE_{10}}$ because an electron is lost from a new shell; it is less shielded from, and closer to, the nucleus; therefore there is a significantly greater force of attraction between the electron and the nucleus;
- there is a steady increase from $\mathbf{IE_2} \longrightarrow \mathbf{IE_9}$ because electrons are being removed from ions with an increasingly positive charge; therefore there is a steadily increasing force of attraction between nucleus and electrons.

A plot of IE against number of electron removed for **$IE_2 \longrightarrow IE_9$** provides evidence of **subshells**.

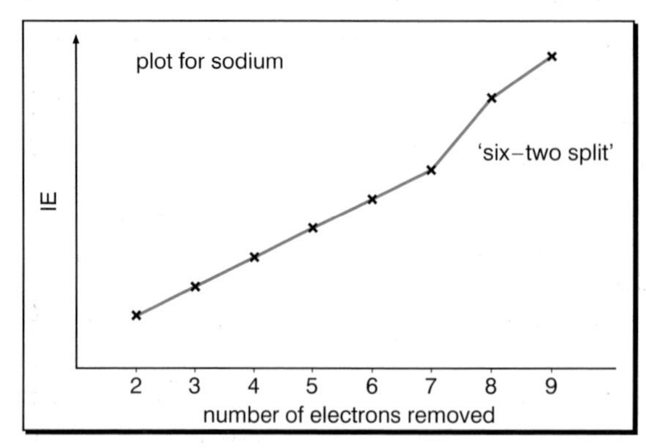

This graph provides evidence that, in sodium, the second quantum shell is divided into two subshells, with two electrons at a lower energy level than the other six.

A more detailed representation of the electron configuration of sodium is: **Na: $1s^2\ 2s^2\ 2p^6\ 3s^1$** (subshell notation) instead of simply **Na: 2, 8, 1** (shell notation).

Orbitals

An **orbital** is a region, called a **probability envelope**, within which there is an approximately 98% chance of finding an electron.

- An **s** subshell contains a spherical s orbital.

- A **p** subshell contains three p orbitals, denoted p_x, p_y and p_z. A p orbital is a 'dumb-bell' shape.
- A **d** subshell contains five d orbitals, all of complex shapes.

Elements in the same period

Example The third period of the periodic table (sodium to argon)

From the graph, note that there is:

- a general increase across a period, because the outermost electron is lost from the same shell and there is the same shielding; the number of protons in the nucleus of each atom is increasing and so the attraction between outer electrons and nucleus increases;

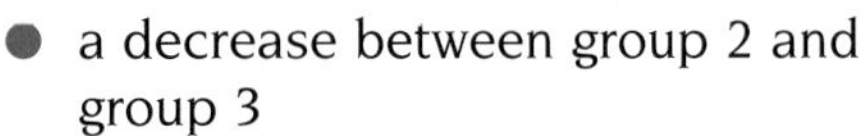

- a decrease between group 2 and group 3

  magnesium: $1s^2\ 2s^2\ 2p^6\ 3s^2$

  aluminium: $1s^2\ 2s^2\ 2p^6\ 3s^2\ 3p^1$

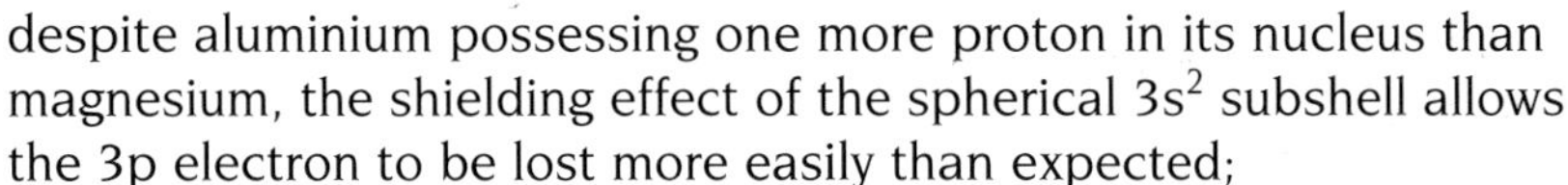

  despite aluminium possessing one more proton in its nucleus than magnesium, the shielding effect of the spherical $3s^2$ subshell allows the 3p electron to be lost more easily than expected;

- a decrease between group 5 and group 6

  phosphorus: $1s^2\ 2s^2\ 2p^6\ 3s^2\ 3p^3$

  sulphur: $1s^2\ 2s^2\ 2p^6\ 3s^2\ 3p^4$

  despite sulphur possessing one more proton in its nucleus than phosphorus, the increase in repulsive forces when **spin-pairing** occurs for the first time in the p subshell enables the electron in sulphur to be lost more easily than an unpaired electron in phosphorus.

The building-up principle

The electronic configurations of isolated atoms, up to Z = 36, can be predicted using the **building-up principle**. There are three important rules to follow:

1 of the orbitals available, the added electron occupies the orbital of lowest energy;

2 each orbital can hold a maximum of two electrons, but they must have opposite spin;

3 if a number of orbitals of equal energy is available, e.g. $2p_x$ $2p_y$ $2p_z$, the added electron will go into a vacant orbital, keeping spins parallel, before two electrons can occupy an orbital (spin-pairing).

Example An isolated carbon atom has the electronic configuration C: $1s^2$ $2s^2$ $2p^2$ or, using the electrons-in-boxes notation:

C:

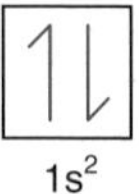

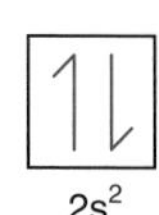

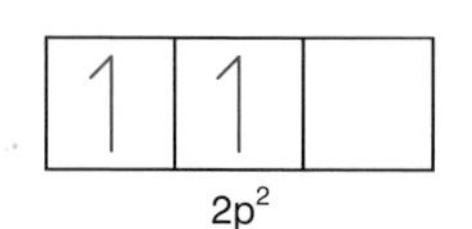

To remember the order of increasing energy of the atomic orbitals, write down the orbitals in the following columns and then draw diagonal lines as shown:

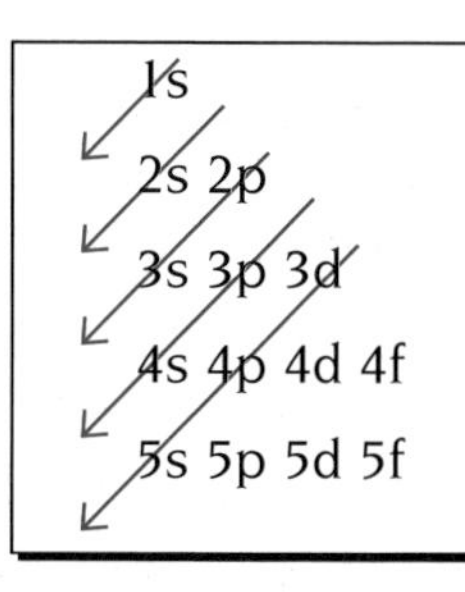

Examples

$_{11}$Na: $1s^2$ $2s^2$ $2p^6$ $3s^1$
$_{17}$Cl: $1s^2$ $2s^2$ $2p^6$ $3s^2$ $3p^5$
$_{20}$Ca: $1s^2$ $2s^2$ $2p^6$ $3s^2$ $3p^6$ $4s^2$
$_{22}$Ti: $1s^2$ $2s^2$ $2p^6$ $3s^2$ $3p^6$ $3d^2$ $4s^2$

note that after Z = 20, the 3d orbital falls below the 4s orbital in energy
$_{36}$Kr: $1s^2$ $2s^2$ $2p^6$ $3s^2$ $3p^6$ $3d^{10}$ $4s^2$ $4p^6$

Electron affinity

For non-metal elements, such as the halogens (group 7), the concept of **electron affinity** is far more useful than ionisation energy. These elements form negative ions in ionic compounds.

First electron affinity is the energy change accompanying the gain of one mole of electrons by one mole of gaseous atoms, to form one mole of singly charged negative ions.

$X_{(g)} + e^- \longrightarrow X^-_{(g)}$

Second electron affinity is the energy change accompanying the gain of one mole of electrons by one mole of singly charged negative ions in the gas phase, to form one mole of double negatively charged ions.

$X^-_{(g)} + e^- \longrightarrow X^{2-}_{(g)}$

The first electron affinity of each halogen in the series chlorine to iodine is shown in the table below.

Cl	-348 kJ mol^{-1}
Br	-324 kJ mol^{-1}
I	-295 kJ mol^{-1}

As the size of the halogen atom increases, the force of attraction between the nucleus of the atom and the incoming electron decreases. Therefore, the first electron affinities become less exothermic down group 7.

Example Second electron affinities

$O^-_{(g)} + e^- \longrightarrow O^{2-}_{(g)}$ is endothermic, as forces of repulsion must be overcome when adding an electron to a negatively charged ion.

Check yourself

1 Atoms are made up of protons, neutrons and electrons. What is the relative mass and charge of each? (3)

2 Define the term mass number. (1)

3 Define the term atomic number. (1)

4 Define the term relative atomic mass. (2)

5 What are isotopes? (2)

6 Naturally occurring magnesium consists of three isotopes, ^{24}Mg, ^{25}Mg and ^{26}Mg, with relative abundances 78.6%, 10.1% and 11.3%, respectively. Calculate the relative atomic mass of magnesium, correct to three significant figures. (3)

7 A mass spectrometer can be used to obtain data needed to calculate relative atomic mass. How are ions (a) accelerated and (b) deflected in a mass spectrometer? (2)

8 Explain why the first ionisation energies of the group 1 elements decrease down the group. (2)

9 Write down in full, using subshell notation, the electronic configuration of (a) the isolated fluorine atom (F has atomic number 9) and (b) the chloride ion Cl^- (Cl has atomic number 17). (2)

10 Magnesium burns exothermically in oxygen:

$Mg_{(s)} + \frac{1}{2}O_{2(g)} \longrightarrow Mg^{2+}\ O^{2-}{}_{(s)}$

Given that the sum of the first two ionisation energies of magnesium is endothermic and that the sum of the first two electron affinities of oxygen is also endothermic, suggest why the reaction between magnesium and oxygen is, nonetheless, exothermic overall. (2)

The answers are on page 105.

The mole concept

Atoms are too small to be seen. Weighing an individual atom, or a small number of atoms, is of little value to us. To obtain a sample of just 1 g of any element in the periodic table, a very large number of atoms would have to be weighed. Instead of grams, a suitable quantity for comparing atoms of elements is the **mole**.

A mole is the amount of a substance that contains as many elementary particles as there are atoms in 12 g of the carbon-12 isotope.

The **Avogadro constant** is the number of particles contained in one mole of a substance. Its value is 6.02×10^{23} mol^{-1} and its symbol is *L*. In one mole of the carbon-12 isotope, there are 6.02×10^{23} ^{12}C atoms.

One mole of atoms of an element is its **relative atomic mass** expressed in grams.

For molecular and ionic substances, one mole of any substance is its **relative molecular mass** or **relative formula mass** expressed in grams. This is called the **molar mass**, i.e. the mass in grams of one mole of a substance.

To a chemist, the **amount** of a substance is measured in **moles** (abbreviated to **mol**). Relative atomic masses (or molar masses in $g\,mol^{-1}$) do not have to be learnt because examination groups always include them on question papers.

Calculation of moles of atoms

$$\text{Number of moles of atoms of an element} = \frac{\text{mass of element}}{\text{molar mass}}$$

This formula can be checked using units of the quantities involved:

$$\text{mol} = \frac{\text{g}}{\text{g mol}^{-1}}$$

Example In 46 g of sodium (relative atomic mass Na = 23)

$$\text{Number of moles of Na} = \frac{46\text{ g}}{23\text{ g mol}^{-1}} = 2\text{ mol}$$

Calculation of moles of molecules

To calculate the relative molecular mass of one mole of molecules of a substance, add up the relative atomic masses of the constituent elements.

Oxygen gas has the formula O_2. As the relative atomic mass of oxygen is 16, the relative molecular mass of O_2 = 2×16 = 32

Or, the molar mass of oxygen O_2 = 32 g mol^{-1}

Example In 96 g of oxygen gas, O_2

Number of moles of O_2 molecules

$$= \frac{\text{mass of substance (element or compound)}}{\text{molar mass}}$$

$$= \frac{96\text{ g}}{32\text{ g mol}^{-1}} = 3\text{ mol}$$

Calculation of moles of a compound

The same formula can be used to calculate the amount of a compound, such as sodium chloride.

Example To calculate the number of moles of sodium chloride in 117 g of the compound:

$$\text{Number of moles of NaCl} = \frac{\text{mass of compound}}{\text{molar mass}}$$

For sodium chloride (relative atomic masses Na = 23.0, Cl = 35.5), the molar mass = 58.5 g mol^{-1} therefore:

$$\text{Number of moles of NaCl} = \frac{117\text{ g}}{58.5\text{ g mol}^{-1}} = 2.00\text{ mol}$$

Empirical formulae

The **empirical formula** is the simplest formula of a compound. It shows the ratio of the elements present in smallest whole numbers. The formula of an ionic compound is always an empirical formula, for example the empirical formula of aluminium oxide is Al_2O_3.

Calculation of empirical formulae

The empirical formula of a compound is determined from reacting mass data.

Example Calculate the empirical formula of the compound containing 6 g of carbon and 1 g of hydrogen (relative atomic masses C = 12, H = 1).

	C : H
1 Mass ratio/g	6 : 1
2 Mole ratio/mol	$\frac{6}{12} : \frac{1}{1}$
=	0.5 : 1
3 Divide by smallest number	$\frac{0.5}{0.5} : \frac{1}{0.5}$
(the simplest mole ratio) =	1 : 2

Answer The empirical formula is CH_2.

Empirical formulae may also be calculated from percentage composition by mass in a similar way.

Example Calculate the empirical formula of the compound that contains 27.30% of carbon and 72.70% of oxygen by mass (relative atomic masses C = 12, O = 16).

	C : O
1 Mass ratio/g	27.30 : 72.70
2 Mole ratio/mol	$\frac{27.30}{12} : \frac{72.70}{16}$
=	2.275 : 4.544

3 Divide by smallest number $\frac{2.275}{2.275} : \frac{4.544}{2.275}$

$= 1 : 2$ (nearest whole number)

Answer The empirical formula is CO_2.

Molecular formulae

The **molecular formula** shows the actual number of each type of atom in a molecule. The molecular formula is a simple multiple of the empirical formula; often the molecular formula is the empirical formula.

Calculation of molecular formulae

To find the molecular formula of a compound, both the empirical formula and the molar mass (in g mol^{-1}) of the compound must be known.

Example A compound of nitrogen and hydrogen has an empirical formula NH_2. The molar mass of the compound is 32 g mol^{-1}. Calculate the molecular formula of the compound (relative atomic masses N = 14, H = 1).

Molar mass of compound = $n \times$ empirical mass of compound

n = an integer (scale factor)

The empirical mass is $(14 \times 1 + 1 \times 2) = 16$ g mol^{-1}

Therefore, to calculate the scale factor, n,

$$n = \frac{\text{molar mass of compound}}{\text{empirical mass of compound}} \qquad n = \frac{32 \text{ g mol}^{-1}}{16 \text{ g mol}^{-1}} \qquad n = 2$$

It follows that the molecular formula of the compound is $(NH_2) \times 2 = N_2H_4$

You can check that this agrees with the data in the question, because:

Molar mass of N_2H_4 in g mol^{-1} $= (14 \times 2) + (1 \times 4)$

$= 32$ g mol^{-1}

Chemical equations

In a **chemical equation**, the substances on the **left-hand side** of the arrow are called the **reactants** and the substances on the **right-hand side** of the arrow are called the **products**.

Example A + B $\longrightarrow$ C + D

A and B are the reactants, whereas C and D are the products.

Equations must **balance** for both **mass** and, where relevant, **total charge**.

Example Balancing for mass

Sodium + water $\longrightarrow$ sodium hydroxide + hydrogen

First, write the correct formulae:

$Na_{(s)} + H_2O_{(l)} \longrightarrow NaOH_{(aq)} + H_{2(g)}$
(unbalanced)

By altering the numbers of each **species**, we obtain a balanced equation:

$2Na_{(s)} + 2H_2O_{(l)} \longrightarrow 2NaOH_{(aq)} + H_{2(g)}$

Left-hand side	Right-hand side
Na = 2	Na = 2
H = 4	H = 4
O = 2	O = 2

Example Balancing for total charge

iron (III) ions + iodide ions $\longrightarrow$ iron (II) ions + iodine molecules

$Fe^{3+}_{(aq)} + 2I^{-}_{(aq)} \longrightarrow Fe^{2+}_{(aq)} + I_{2(aq)}$

This equation is balanced for mass, but unbalanced for charge. Doubling the moles of iron (III) and iron (II) ions balances the equation for total charge:

$2Fe^{3+}_{(aq)} + 2I^{-}_{(aq)} \longrightarrow 2Fe^{2+}_{(aq)} + I_{2(aq)}$

Total charge on left	Total charge on right
$6^{+} + 2^{-} = 4^{+}$	4^{+}

Ionic equations

Ionic equations provide useful 'summaries' of the overall changes occurring in a chemical reaction. In an ionic equation, only the species taking part in the reaction are included.

To convert a full equation into an ionic equation, three steps are taken.

1. For soluble ionic compounds, ions are written separately using the state symbol (aq).
2. Covalent substances and ionic precipitates are written in full.
3. Ions that appear on both sides of the equation and that do not take part in the reaction (**spectator ions**) are deleted.

An example is the **precipitation reaction** between aqueous barium chloride and aqueous sodium sulphate. The full equation is:

$BaCl_{2(aq)} + Na_2SO_{4(aq)} \longrightarrow BaSO_{4(s)} + 2NaCl_{(aq)}$

Combining steps 1 and 2:

$Ba^{2+}{}_{(aq)} + 2Cl^{-}{}_{(aq)} + 2Na^{+}{}_{(aq)} + SO_4{}^{2-}{}_{(aq)} \longrightarrow BaSO_{4(s)} + 2Na^{+}{}_{(aq)} + 2Cl^{-}{}_{(aq)}$

Step 3, deletion of spectator ions, produces the ionic equation:

$Ba^{2+}{}_{(aq)} + SO_4{}^{2-}{}_{(aq)} \longrightarrow BaSO_{4(s)}$

This equation, therefore, represents the reaction between any aqueous solution containing barium ions and any aqueous solution containing sulphate ions. This ionic equation summarises the test for a sulphate ion in aqueous solution.

Using chemical equations

Chemical equations can be used to **work out reacting masses**. A balanced equation tells us the numbers of moles of each reactant that are required to produce the expected amount(s) of product in a chemical reaction.

Example $2H_{2(g)} + O_{2(g)} \longrightarrow 2H_2O_{(l)}$ tells us that 2 moles of $H_{2(g)}$ molecules react with 1 mole of $O_{2(g)}$ molecules $\longrightarrow$ 2 moles of $H_2O_{(l)}$ molecules.

Questions are often set where scaling of quantities is required.

Example Calculate the mass of calcium oxide obtained when 20.0 g of calcium carbonate are thermally decomposed (relative atomic masses Ca = 40, C = 12, O = 16).

$CaCO_{3(s)} \longrightarrow CaO_{(s)} + CO_{2(g)}$

1 mole of $CaCO_{3(s)} \longrightarrow$ 1 mole of $CaO_{(s)}$ + 1 mole of $CO_{2(g)}$

100 g of $CaCO_{3(s)} \longrightarrow$ 56.0 g of $CaO_{(s)}$ + 44.0 g of $CO_{2(g)}$

$\downarrow \div 100$ $\qquad$ $\downarrow \div 100$

$\frac{100}{100} = 1$ g $CaCO_{3(s)} \longrightarrow \frac{56.0}{100}$ g $CaO_{(s)}$

$\downarrow \times 20.0$ $\qquad$ $\downarrow \times 20.0$

20.0 g $CaCO_{3(s)} \longrightarrow \frac{20.0 \times 56.0}{100}$ g $CaO_{(s)}$

$= 11.2$ g $CaO_{(s)}$

Answer 11.2 g of $CaO_{(s)}$ are produced.

Using reacting masses

Reacting masses can be used to **work out chemical equations**. If the mass of each substance taking part in a chemical reaction is known, the balanced chemical equation can then be constructed.

Example In an experiment, 5.60 g of iron combined with 10.65 g of chlorine gas to form a chloride of iron. Construct the equation for the reaction (relative atomic masses Fe = 56, Cl = 35.5).

$$\text{Moles of } Fe_{(s)} = \frac{\text{mass of } Fe_{(s)}}{\text{molar mass}}$$

$$= \frac{5.60 \text{ g}}{56 \text{ g mol}^{-1}}$$

$$= 0.10 \text{ mol Fe}$$

Moles of $Cl_{2(g)}$ $= \frac{\text{mass of } Cl_{2(g)}}{\text{molar mass}}$

$= \frac{10.65 \text{ g}}{71 \text{ g mol}^{-1}}$

$= 0.15 \text{ mol } Cl_2$

Therefore	Fe	+	Cl_2
Mole ratio/mol	0.10	:	0.15
Divide by smallest number	$\frac{0.10}{0.10}$	:	$\frac{0.15}{0.10}$
	= 1	:	1.5
	= 2	:	3

Answer $2Fe_{(s)} + 3Cl_{2(g)} \longrightarrow 2FeCl_{3(s)}$

iron + chlorine $\longrightarrow$ iron (III) chloride

Gas volumes

It is often far more convenient to measure the **volume** of a gas rather than its **mass**. Avogadro's hypothesis states that 'equal volumes of all gases contain the same number of particles at the same temperature and pressure'. The **molar volume** of a gas is taken as 24 dm^3 (24,000 cm^3) at room temperature and pressure (i.e. the volume of one mole of any gas).

Example Calculate the volume of carbon dioxide obtained, at room temperature and pressure, when 25 g of calcium carbonate undergo thermal decomposition (relative atomic masses Ca = 40, C = 12, O = 16).

$CaCO_{3(s)} \longrightarrow CaO_{(s)} + CO_{2(g)}$

From the equation

$1 \text{ mole } CaCO_{3(s)} \longrightarrow 1 \text{ mole } CaO_{(s)} + 1 \text{ mole } CO_{2(g)}$

Therefore

$$
\begin{array}{lll}
100 \text{ g } CaCO_{3(s)} & \longrightarrow & 56 \text{ g } CaO_{(s)} + 24 \text{ dm}^3\ CO_{2(g)} \\
\downarrow \div 100 & & \downarrow \div 100 \\
1.0 \text{ g } CaCO_{3(s)} & \longrightarrow & \frac{24}{100} \text{ dm}^3\ CO_{2(g)} \\
\downarrow \times 25 & & \downarrow \times 25 \\
25 \text{ g } CaCO_{3(s)} & \longrightarrow & \frac{25 \times 24}{100} \text{ dm}^3\ CO_{2(g)} \\
 & & = 6.0 \text{ dm}^3\ CO_{2(g)}
\end{array}
$$

Answer 6.0 dm^3 $CO_{2(g)}$ are produced, measured at room temperature and pressure.

Reacting gas volumes

In order to work out **reacting volumes of gases**, we can use **Gay–Lussac's law** of combining volumes: 'When gases react they do so in volumes which bear a simple ratio to each other and to the volumes of the products if they are gases, all measurements of volume being at the same temperature and pressure'.

Example Methane burns in oxygen to give carbon dioxide and water

$CH_{4(g)} + 2O_{2(g)} \longrightarrow CO_{2(g)} + 2H_2O_{(l)}$

(a) Calculate the volume of oxygen needed to burn 10 dm^3 of methane.

(b) Calculate the volume of carbon dioxide produced when 10 dm^3 of methane are burned.

All gas volumes are measured at the same temperature and pressure.

From the equation

$CH_{4(g)} + 2O_{2(g)} \longrightarrow CO_{2(g)} + 2H_2O_{(l)}$

$1 \text{ mole } CH_{4(g)} : 2 \text{ moles } O_{2(g)} \longrightarrow 1 \text{ mole } CO_{2(g)}$

Therefore 1 volume of $CH_{4(g)}$: 2 volumes of $O_{2(g)}$ $\longrightarrow$ 1 volume of $CO_{2(g)}$

Therefore 10 dm^3 of $CH_{4(g)}$: 20 dm^3 of $O_{2(g)}$ $\longrightarrow$ 10 dm^3 of $CO_{2(g)}$

Answer (a) 20 dm^3 of oxygen are required to completely burn 10 dm^3 of methane. (b) 10 dm^3 of carbon dioxide are produced when 10 dm^3 of methane are completely burned.

Volumetric analysis

The concentration of a solution can be determined using a procedure known as **titration**. A carefully measured volume of the solution being analysed is reacted completely with another solution of accurately known concentration and volume. The concentration of the former solution can then be found, from a knowledge of the reaction occurring between the two substances.

Concentration of solutions

A solution containing 1 mole of solute in 1 dm^3 (1000 cm^3) of solution is called a 1 molar solution or a solution of concentration 1 mol dm^{-3}.

$$\text{Concentration} = \frac{\text{amount of solute (mol)}}{\text{volume of solution (dm}^3\text{)}}$$

Units of concentration: mol dm^{-3}

Alternatively, the concentration of a solution can be expressed in grams per dm^3, in which case:

$$\text{Concentration} = \frac{\text{mass of solute (g)}}{\text{volume of solution (dm}^3\text{)}}$$

Units of concentration: g dm^{-3}

Calculating numbers of moles

The numbers of moles can be calculated from the concentration and volume of a solution.

Example Calculate the number of moles of NaOH present in 40 cm^3 of a 2 mol dm^{-3} solution of sodium hydroxide.

From the definition of concentration, there are 2 mol NaOH per 1 dm^3 solution

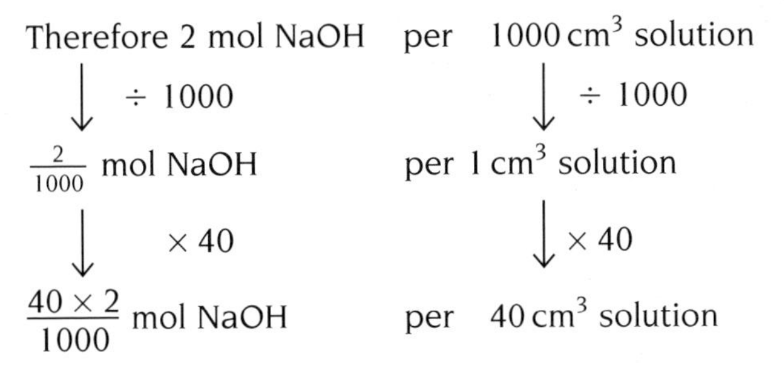

Answer 0.08 mol NaOH per 40 cm^3 solution

In general: Amount of solute (mol)

$$= \frac{\text{concentration of solution (mol dm}^{-3}) \times \text{volume of solution (cm}^3)}{1000}$$

Units (mol) = (mol dm^{-3}) × (dm^3)

Acid–base titrations

When an acid and a base react together, an **indicator** can be used to show when the reaction is just complete, called the **equivalence point**.

Example Hydrochloric acid of concentration 0.100 mol dm^{-3} was added from a burette into a flask containing 25.0 cm^3 of aqueous sodium hydroxide until neutralisation took place. 20.0 cm^3 of the acid were required. Calculate the concentration of the sodium hydroxide solution in mol dm^{-3}.

$$NaOH_{(aq)} + HCl_{(aq)} \longrightarrow NaCl_{(aq)} + H_2O_{(l)}$$

$$\text{Moles of HCl}_{(aq)}\text{ reacting} = \frac{\text{concentration HCl} \times \text{volume HCl}}{1000}$$

$$= \frac{0.100 \times 20.0}{1000}$$

$$= 0.00200 \text{ moles HCl}_{(aq)}$$

From the balanced equation, it can be seen that 0.00200 moles $NaOH_{(aq)}$ are required.

$$\text{Moles of NaOH}_{(aq)}\text{ reacting} = \frac{\text{concentration of NaOH (mol dm}^{-3}) \times \text{volume of NaOH (cm}^3)}{1000}$$

Rearranging the above equation:

$$\text{Concentration of NaOH}_{(aq)} = \frac{\text{moles of NaOH}_{(aq)}\text{ reacting} \times 1000}{\text{volume of NaOH}_{(aq)}}$$

$$= \frac{0.00200 \times 1000}{25.0}$$

$$= 0.0800 \text{ mol dm}^{-3}$$

Answer The concentration of the sodium hydroxide solution is $0.0800 \text{ mol dm}^{-3}$

Yield calculations

Many chemical reactions do not run to completion. Particularly in organic chemistry, side reactions can occur and product can be lost during purification procedures. The **yield** of product will, therefore, be less than 100%. The **percentage yield** can be calculated from the following equation:

$$\text{Percentage yield} = \frac{\text{actual mass of product}}{\text{calculated mass of product}} \times 100$$

Example On oxidation of 50.0 g of ethanol, 59.0 g of ethanoic acid were obtained. Calculate the percentage yield of the product (molar masses ethanol $C_2H_5OH = 46.0\,g\,mol^{-1}$, ethanoic acid $CH_3CO_2H = 60.0\,g\,mol^{-1}$).

$C_2H_5OH + 2[O] \longrightarrow CH_3CO_2H + H_2O$
ethanol → ethanoic acid

From the equation

1 mole $C_2H_5OH \longrightarrow$ 1 mole CH_3CO_2H

46.0 g $C_2H_5OH \longrightarrow$ 60.0 g CH_3CO_2H

1.00 g C_2H_5OH should give $\frac{60.0}{46.0}$ g CH_3CO_2H

50.0 g C_2H_5OH should give $\frac{50.0 \times 60.0}{46.0}$ g CH_3CO_2H

50.0 g C_2H_5OH should give 65.2 g CH_3CO_2H

$$\text{Therefore, percentage yield} = \frac{\text{actual mass of product}}{\text{calculated mass of product}} \times 100$$

$$= \frac{59.0}{65.2} \times 100$$

Answer 90.5%

Check yourself

1 In this question, assume that the value of the Avogadro constant, L, is 6.0×10^{23} mol^{-1}.

(a) Calculate the number of carbon atoms, C, in 0.120 g of carbon (relative atomic mass C = 12.0). (2)

(b) Calculate the number of sulphur dioxide molecules, SO_2, in 32.0 g of sulphur dioxide (relative atomic masses S = 32, O = 16). (2)

(c) Calculate the number of sodium ions, Na^+, in 14.2 g of sodium sulphate (relative atomic masses Na = 23, S = 32, O = 16). (2)

2 Calculate the number of moles of atoms present in each of the following (relative atomic masses are given in brackets):

(a) 16.0 g of oxygen atoms (O = 16.0) (2)

(b) 0.140 g of nitrogen atoms (N = 14.0) (2)

(c) 5.40 g of silver atoms (Ag = 108) (2)

3 Calculate the mass, in grams, of each of the following amounts (relative atomic masses are given in brackets):

(a) 0.500 mol of oxygen atoms (O = 16.0) (1)

(b) 10.0 mol of sodium atoms (Na = 23.0) (1)

(c) 0.0100 mol of hydrogen atoms (H = 1) (1)

4 Calculate the molar masses, in g mol^{-1}, of each of the following substances:

(a) Br_2 (Br = 80) (1)

(b) HNO_3 (H = 1, N = 14, O = 16) (1)

(c) $CuSO_4.5H_2O$ (Cu = 64, S = 32, O = 16, H = 1) (1)

5 Calculate the number of moles present in each of the following samples:

(a) 128 g of oxygen, O_2 (O = 16.0) (1)

(b) 25.25 g of potassium nitrate, KNO_3 (K = 39, N = 14, O = 16) (1)

(c) 414 g of ethanol, C_2H_5OH (C = 12, H = 1, O = 16) (1)

6 Calculate the mass, in grams, of each of the following amounts:

(a) 2.00 mol of sulphur dioxide molecules, SO_2 (S = 32, O = 16) (1)

The answers are on pages 106 and 107.

(b) 20.0 mol of sulphuric acid, H_2SO_4 (H = 1, S = 32, O = 16) (1)

(c) 0.500 mol of sodium hydroxide, NaOH (Na = 23, O = 16, H = 1) (1)

7 Calculate the empirical formulae of the following compounds from the information given:

(a) 1.12 g of iron combines with oxygen to form 1.60 g of an oxide of iron (Fe = 56, O = 16) (2)

(b) a compound was found to contain 43.4% sodium, 11.3% carbon and 45.3% oxygen by mass (Na = 23.0, C = 12.0, O = 16.0) (2)

(c) a compound contains 82.75% by mass of carbon and 17.25% by mass of hydrogen (C = 12, H = 1) (2)

8 Work out the molecular formula of the compound determined in Question 7(c) above, given that its molar mass is 58 $g\,mol^{-1}$. (2)

9 Write ionic equations derived from each of the following full chemical equations:

(a) $AgNO_{3(aq)} + NaCl_{(aq)} \longrightarrow AgCl_{(s)} + NaNO_{3(aq)}$ (2)

(b) $Zn_{(s)} + H_2SO_{4(aq)} \longrightarrow ZnSO_{4(aq)} + H_{2(g)}$ (2)

(c) $HCl_{(aq)} + NaOH_{(aq)} \longrightarrow NaCl_{(aq)} + H_2O_{(l)}$ (2)

10 Calculate the mass of magnesium chloride that would be obtained when 48 g of magnesium metal is reacted completely with chlorine (Mg = 24, Cl = 35.5).

$Mg_{(s)} + Cl_{2(g)} \longrightarrow MgCl_{2(s)}$ (2)

11 Calculate the mass of calcium carbonate that must be decomposed in order to produce 14 tonnes of calcium oxide, CaO (Ca = 40, C = 12, O = 16).

$CaCO_{3(s)} \longrightarrow CaO_{(s)} + CO_{2(g)}$ (2)

12 9.80 g of sulphuric acid reacted with 8.50 g of sodium nitrate to produce 6.30 g of nitric acid. Use the information to work out the equation for the reaction, given that there are only two products of the reaction (molar masses H_2SO_4 = 98 $g\,mol^{-1}$, $NaNO_3$ = 85 $g\,mol^{-1}$ and HNO_3 = 63 $g\,mol^{-1}$). (3)

The answers are on page 107.

Check yourself

13 Calculate the volume of hydrogen gas, measured at room temperature and pressure, which would be formed when 12 g of magnesium react with excess dilute sulphuric acid (assume that one mole of any gas occupies a volume of 24,000 cm^3 (24 dm^3) at room temperature and pressure; relative atomic mass of Mg = 24). (2)

$Mg_{(s)} + H_2SO_{4(aq)} \longrightarrow MgSO_{4(aq)} + H_{2(g)}$

14 A car engine burns the compound octane C_8H_{18}

$2C_8H_{18(g)} + 25O_{2(g)} \longrightarrow 16CO_{2(g)} + 18H_2O_{(l)}$ Calculate:

(a) The volume of oxygen required to completely burn 2.00 dm^3 of octane vapour. (1)

(b) The volume of carbon dioxide gas you would expect to be formed. (1) (Assume that all gas volumes are measured at the same temperature and pressure.)

15 In a titration, it was found that 28.5 cm^3 of dilute nitric acid (concentration 0.100 mol dm^{-3}) were exactly neutralised by 25.0 cm^3 of potassium hydroxide solution.

$KOH_{(aq)} + HNO_{3(aq)} \longrightarrow KNO_{3(aq)} + H_2O_{(l)}$

Calculate the concentration of the potassium hydroxide solution in mol dm^{-3}. (2)

16 40.0 cm^3 of an aqueous solution of potassium hydroxide containing 5.6 g dm^{-3} are neutralised exactly by 40.0 cm^3 of dilute hydrochloric acid:

$KOH_{(aq)} + HCl_{(aq)} \longrightarrow KCl_{(aq)} + H_2O_{(l)}$

Calculate the concentration of the hydrochloric acid solution in mol dm^{-3} (relative atomic masses K = 39, O = 16, H = 1) (3)

17 In an organic chemistry experiment, 49.2 g of nitrobenzene ($C_6H_5NO_2$) were obtained from 39.0 g of benzene (C_6H_6). The equation for the reaction is:

$C_6H_{6(l)} + HNO_{3(l)} \longrightarrow C_6H_5NO_{2(l)} + H_2O_{(l)}$

Calculate the percentage yield of the reaction (relative atomic masses C = 12, H = 1, N = 14, O = 16). (2)

The answers are on page 110.

There are three types of strong chemical bonding:

- ionic;
- covalent;
- metallic.

Ionic bonding

Ionic bonding is the electrostatic force of attraction between oppositely charged ions (+ and –), which are formed as a result of electron transfer between atoms.

The attractions between positive ions (cations) and negative ions (anions) are strong, and ionic compounds have a giant lattice structure.

Example Sodium chloride Na^+Cl^-

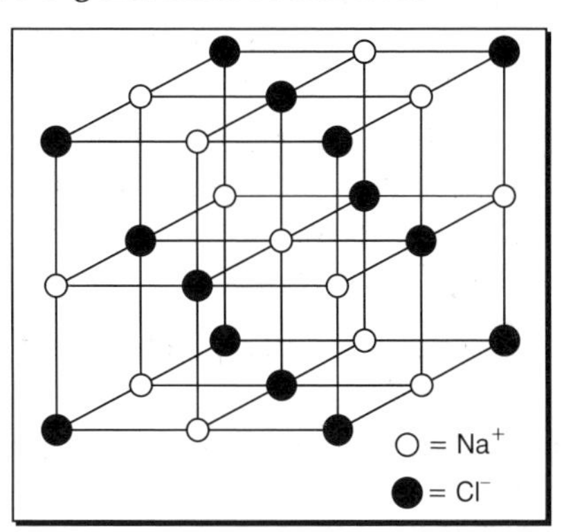

The 3-D arrangement of ions is regular, with alternate positive and negative ions. The lattice structure is one example of the 3-D arrangement.

An ionic bond is normally formed between two elements of very different **electronegativity** (typically greater than 1.5).

Covalent bonding

Covalent bonding is the electrostatic force of attraction of two nuclei each with a shared pair of electrons between them.

In a covalent bond, each atom provides one electron. Overlap of an orbital containing an electron from one atom overlaps with another orbital containing an electron from the second atom.

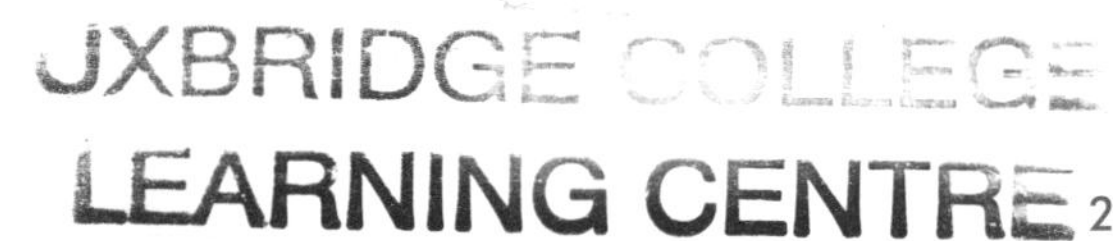

A **sigma (σ) bond** is formed when the overlap occurs along the line between the two nuclei.

Example

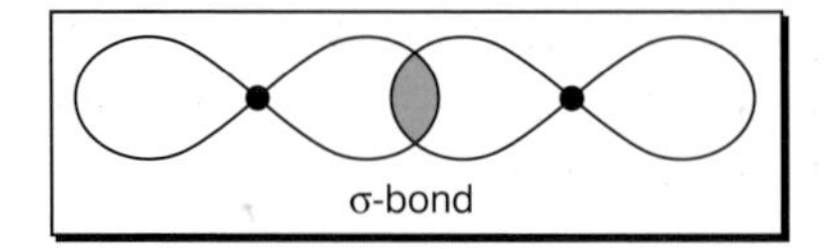

A **pi (π) bond** is formed when the overlap between the two orbitals occurs sideways.

Example In a π bond, there is overlap above and below the line between the two nuclei.

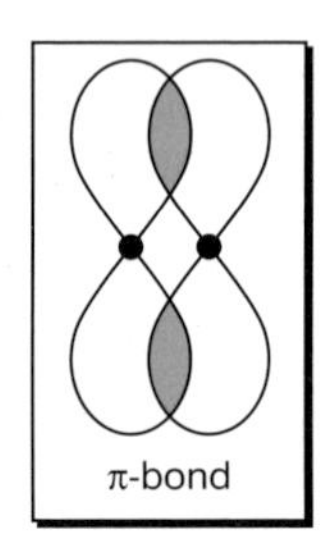

Giant atomic (or **giant molecular**) **structures**, e.g. diamond and graphite, consist of many atoms joined together by very strong covalent bonds. A typical covalent bond is about the same strength as an ionic bond.

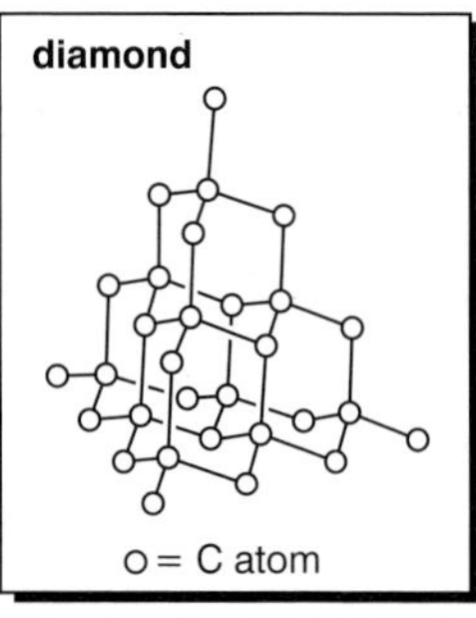

Example Diamond

Each C atom is surrounded by four others, in a tetrahedral 3-D arrangement. The diagram does not show a molecule, but the pattern of arrangement, which continues on and on.

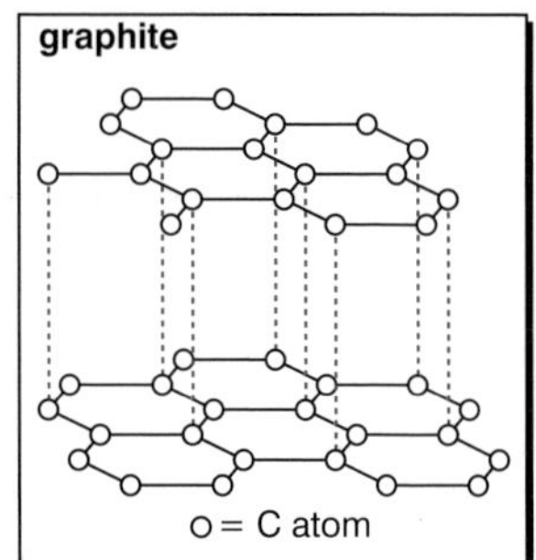

Example Graphite

In graphite, each C atom is bonded covalently to three others. The C atoms are arranged in hexagons within a layer structure. Every C atom has four outer electrons, but only three are involved in covalent bonding. The non-bonded electrons are delocalised and flow along, but not up or down between, the layers. The layers are held together by weak **van der Waals' forces**.

Simple molecules are made up of small groups of atoms, covalently bonded together in a molecule. Although the covalent bonds between the atoms are very strong, there are only weak forces between the molecules.

Example Iodine, I_2

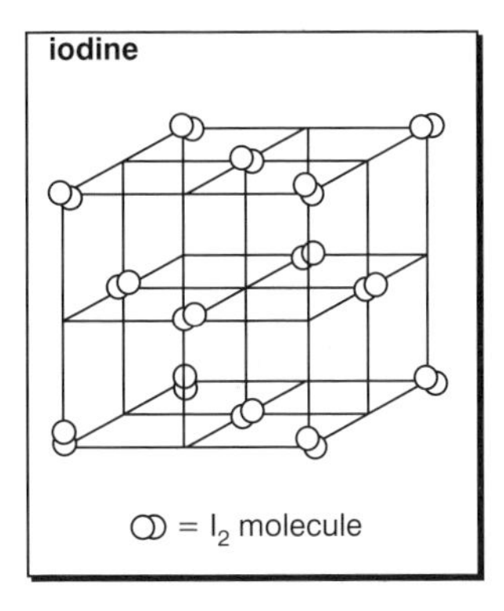

Dative covalent bonding

A **dative covalent bond**, or **co-ordinate bond**, is a covalent bond in which both of the shared pair of electrons come from the same atom.

Examples The following compounds and species are examples of dative covalent bonding.

$BF_3.NH_3$

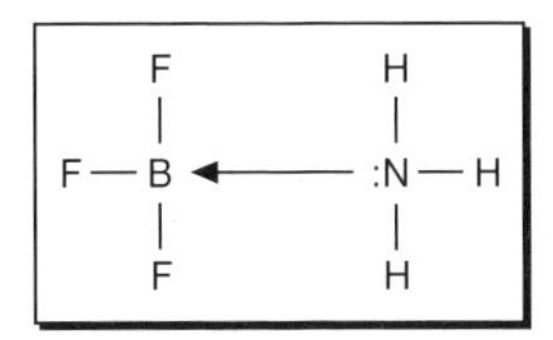

$H_3O^+_{(aq)}$ ion (responsible for acidity in aqueous solution)

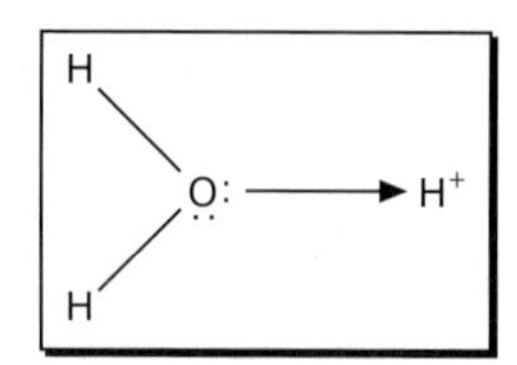

NH_4^+ ion

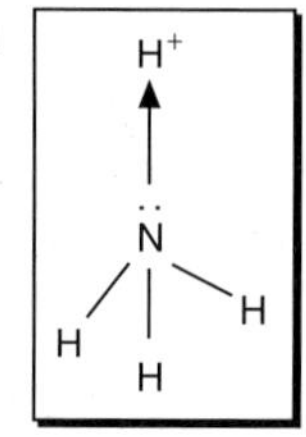

Hydrated metal cations such as $[Mn(H_2O)_6)]^{2+}_{(aq)}$

The water molecules are bonded to the central Mn^{2+} cation via dative covalent bonds from the lone pair on the oxygen atom into empty orbitals in the metal ion.

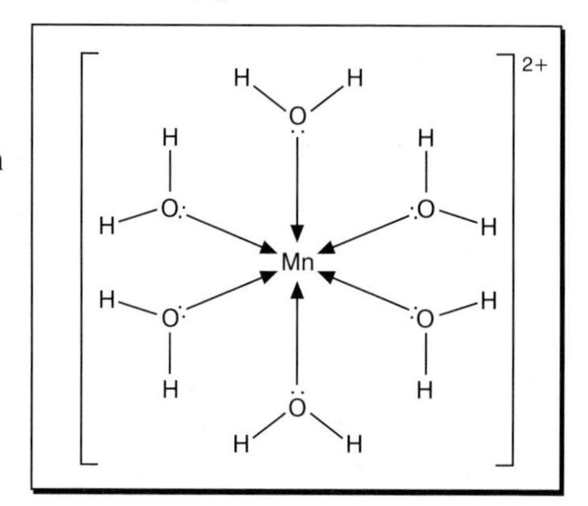

Metallic bonding

Metallic bonding is the electrostatic force of attraction between positively charged metal cations and the delocalised electrons between them.

Example A metal $M_{(s)}$

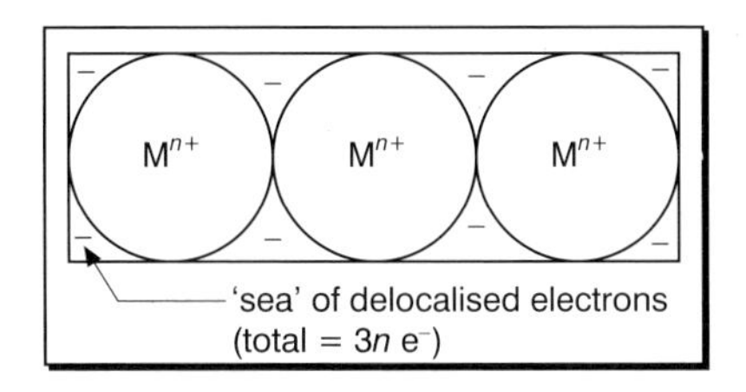

Partial ionic/covalent bonding

The terms 'covalent' and 'ionic' bonding represent the extremes of the bond types. In reality, the bonding in most compounds is intermediate between the two extremes.

Pure ionic			**Pure covalent**
↑ Total transfer of electrons	Partial transfer of electrons	Unequal sharing of electrons	Even sharing of electrons ↑

Pure ionic bonding is favoured in reactions between metal atoms having low ionisation energies and non-metal atoms with highly exothermic electron affinities. In pure ionic compounds, the metal cations will be large ions of low charge and the non-metal anions will be small ions of low charge.

In a pure ionic compound, the electron distribution around the ions is spherical.

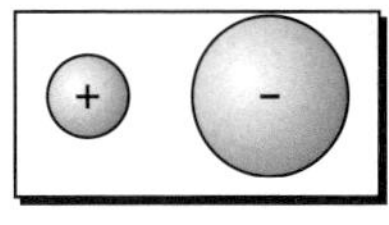

If the cation is small and has a high charge, i.e. it has a high charge density, it is very **polarising**. Such a cation will distort the electron cloud of the anion, pulling the electron density towards itself.

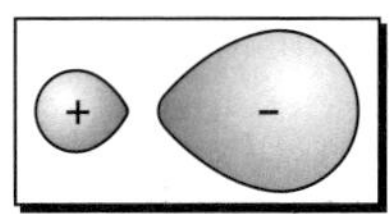

This introduces covalent character into the ionic compound. Anions with a large radius and high negative charge are very polarisable.

In a **purely covalent molecule**, the bonding pair of electrons is evenly shared between the two nuclei.

Example Molecules consisting of two identical atoms, such as H_2 and Cl_2

The electron cloud is equally distributed between the two nuclei.

In molecules where two different atoms are joined by a covalent bond, the bond pair of electrons is not shared equally.

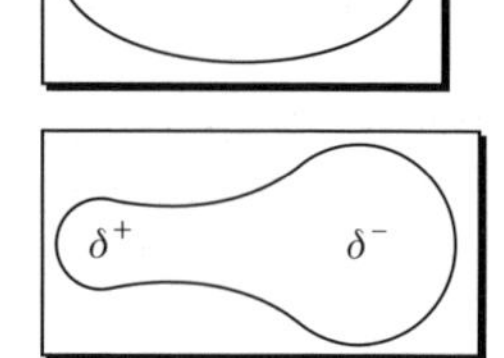

Example Hydrogen chloride molecule

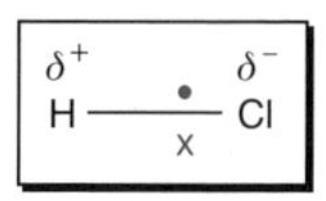

Because the Cl atom is more electronegative than the H atom, the bond pair is pulled towards the Cl atom. This results in a covalent bond with some partial ionic character; the small charges present are indicated by a δ.

The **electronegativity** of an atom is the power of that atom to attract the bonding electrons in a covalent bond.

Electronegativity values, on the **Pauling scale**, of some elements are listed in the table.

Element	Electronegativity
F	4.0
O	3.5
Cl	3.0
N	3.0
C	2.5
H	2.1

Intermolecular forces

Intermolecular forces exist **between** covalent molecules in the solid and liquid states. There are two categories:
- van der Waals' forces;
- hydrogen bonds.

van der Waals' forces

Induced-dipole/induced-dipole forces

Induced-dipole/induced-dipole forces, also called **temporary dipole/temporary dipole forces**, are weak attractive forces that exist between **all** molecules. They arise when an instantaneous imbalance in the electron distribution in a molecule induces a corresponding imbalance in neighbouring molecules, leading to a weak electrostatic attraction.

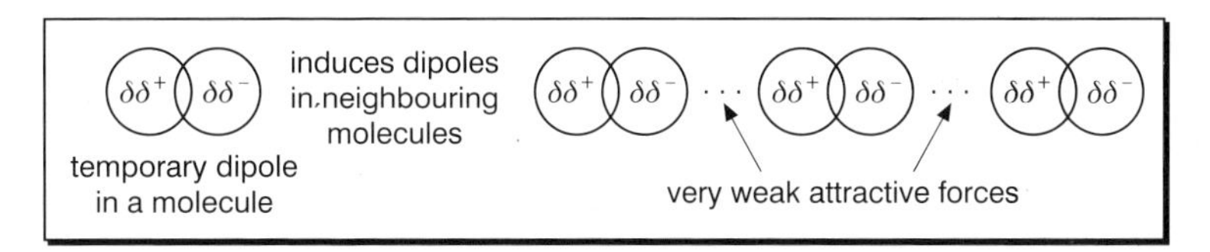

Induced-dipole/induced-dipole forces increase in strength as the number of electrons in the molecule increases. This phenomenon is illustrated by the increase in boiling temperatures of the halogens down group 7.

Permanent-dipole/permanent-dipole forces

Permanent-dipole/permanent-dipole forces are weak attractive forces between permanently polar molecules. δ^+ atoms in one molecule attract δ^- atoms in another molecule. They act in addition to the induced-dipole/induced-dipole forces.

Example Propanone (CH_3COCH_3) and butane ($CH_3CH_2CH_2CH_3$) both have molar masses of 58 g mol^{-1}

The boiling temperature of propanone, however, is 56°C, whereas that of butane is 0°C. This difference is explained by the additional presence of permanent-dipole/permanent-dipole forces between propanone molecules.

$H_3C—C^{\delta+}—CH_3$
‖
$O^{\delta-}$
⋮
$H_3C—C^{\delta+}—CH_3$
‖
$O^{\delta-}$

Hydrogen bonding

The **hydrogen bond** is the strongest intermolecular force, typically about 1/10th the strength of a covalent bond.

A hydrogen bond is a **weak directional bond** formed between a δ^+ hydrogen atom in one molecule and the lone pair of electrons on a highly electronegative atom in another molecule. The highly electronegative atoms referred to are fluorine, oxygen and nitrogen.

The anomalously high boiling points of the hydrides NH_3, H_2O and HF, compared with the other hydrides in groups 5, 6 and 7, are explained by the presence of hydrogen bonds.

The boiling temperatures of the hydrides of group 6 are shown opposite.

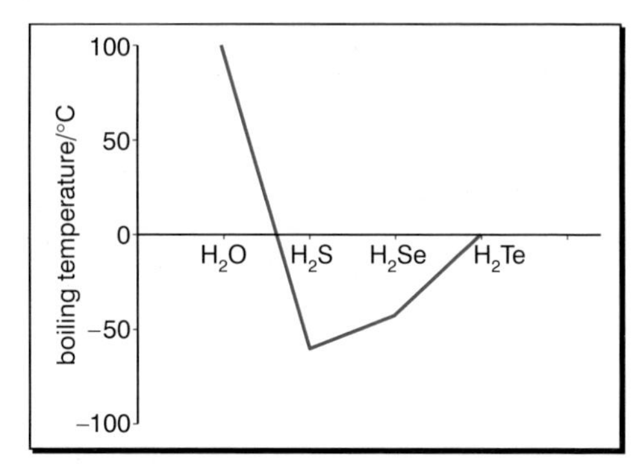

The structure of ice is shown in the diagram. The crystal structure of ice is essentially tetrahedral. When water melts, the hydrogen bonds are progressively broken. The molecules pack closer together and so an initial reduction in volume of the liquid occurs before the usual expansion effect from raising the temperature is observed. Water, therefore, has its maximum density at 4 °C.

Shapes of molecules and ions

The shapes of molecules and ions are explained by the **valence shell electron pair repulsion theory (VSEPR theory)**.

- The electron pairs, whether bonding or lone pairs, arrange themselves around a central atom as far apart as possible. This minimises the **electrostatic repulsion** between the electron pairs.
- The repulsion between lone pairs is greater than that between a lone pair and a bond pair, which is greater than that between bond pairs.
- The electrons in a multiple bond (double or triple) are counted as a single pair when working out the shape of a molecule or ion.

Species without lone pairs

Examples

Two bond pairs, $BeCl_2$

The electrons are at 180°, i.e. it is a linear molecule.

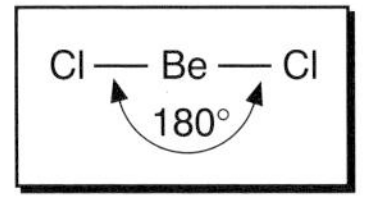

Three bond pairs, BCl_3

The electron pairs are at 120°, i.e. it is a trigonal planar molecule.

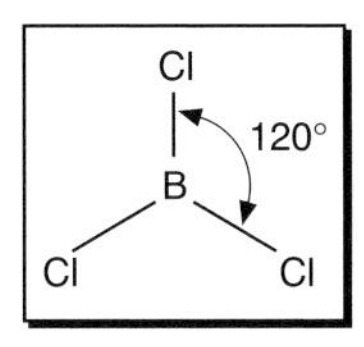

Four bond pairs, CH_4

The electron pairs adopt a tetrahedral arrangement.

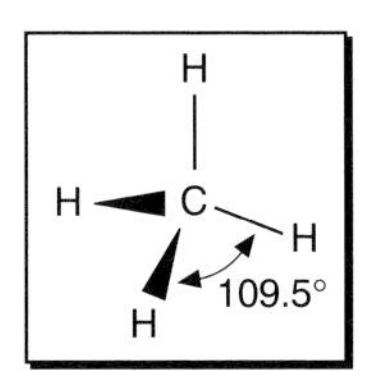

Five bond pairs, PCl_5

The electron pairs are arranged in a trigonal bipyramidal shape.

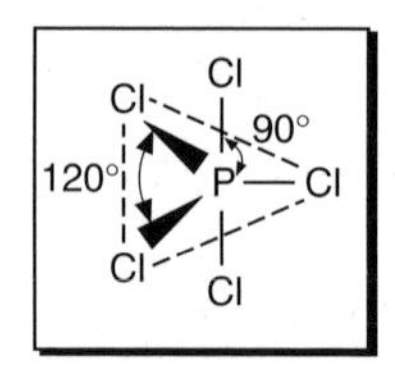

Six bond pairs, SF_6

The electron pairs are arranged in an octahedral shape.

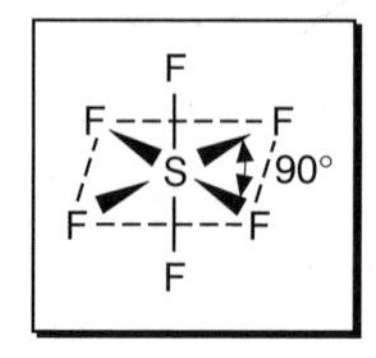

Species with lone pairs

Examples

Four electron pairs as three bond pairs and one lone pair, NH_3 or PCl_3

The shape is a trigonal pyramidal.

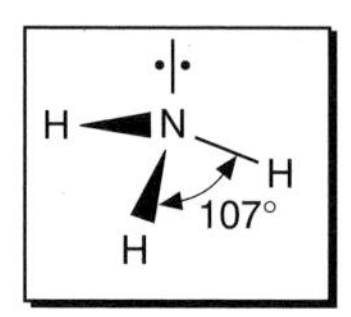

Four electron pairs as two bond pairs and two lone pairs, H_2O

The shape is V-shaped (bent).

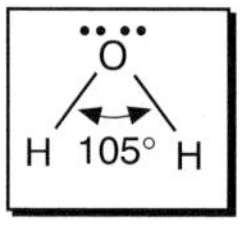

Six electron pairs as four bond pairs and two lone pairs, XeF_4

The shape is square planar.

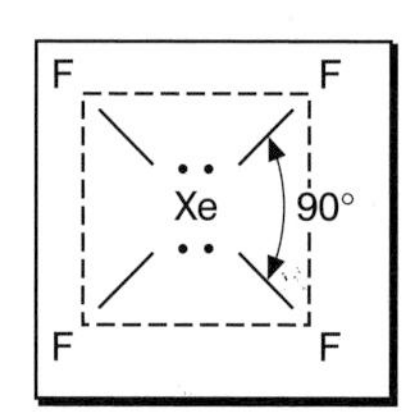

Check yourself

1 Explain why the melting temperature of diamond is much higher that that of iodine. (4)

2 Why does sodium chloride have a relatively high melting point? (2)

3 Explain why aluminium is a good conductor of electricity. (3)

4 Why does magnesium iodide have more covalent character than magnesium chloride? (3)

5 Define the term 'electronegativity'. (2)

6 State the type of structure of each of the following substances in the solid state:

(a) SiO_2 (1)
(b) CO_2 (1)
(c) KBr (1)
(d) Cu (1)

7 The boiling temperatures of the group 7 hydrides, HF to HI, are shown in the table.

Hydride	HF	HCl	HBr	HI
Molar mass/g mol^{-1}	20	36.5	81	128
Boiling temp. / °C	20	−85	−67	−35

(a) Explain why the boiling temperature of HF is higher than that of HCl. (2)
(b) Explain the increase in boiling temperatures from HCl to HI. (2)

8 Use the VSEPR theory to work out the shapes of the following molecules or ions.

(a) SF_6 (2)
(b) CO_3^{2-} (2)
(c) PH_3 (2)
(d) PF_4^+ (2)

The answers are on page 111.

Oxidation and reduction

Oxidation has been defined as:

- gain of oxygen

Example $2Cu_{(s)} + O_{2(g)} \longrightarrow 2CuO_{(s)}$
copper has been oxidised

- loss of hydrogen

Example $MnO_{2(s)} + 4HCl_{(l)} \longrightarrow MnCl_{2(aq)} + 2H_2O_{(l)} + Cl_{2(g)}$
hydrochloric acid has been oxidised

Reduction has been defined as:

- loss of oxygen

Example $Fe_2O_{3(s)} + 3CO_{(g)} \longrightarrow 2Fe_{(l)} + 3CO_{2(g)}$
iron (III) oxide has been reduced

- gain of hydrogen

Example $C_2H_{4(g)} + H_{2(g)} \longrightarrow C_2H_{6(g)}$
ethene has been reduced

Because many reactions involving oxidation and reduction do not involve loss or gain of hydrogen and/or oxygen, it is more useful to define oxidation and reduction in terms of electrons.
Therefore: oxidation is loss of electrons

whereas: reduction is gain of electrons

The phrase **OIL RIG** can be used to help remember this definition. Because **red**uction and **ox**idation occur in the same reaction, such processes are called **REDOX** reactions.

Oxidising agents or **oxidants** are substances that oxidise other reagents. An oxidising agent is, itself, reduced during a reaction. It **gains electrons. Reducing agents** or **reductants** are substances that reduce other reagents. A reducing agent is, itself, oxidised during a reaction. It **loses electrons**.

Reduction and oxidation also take place at electrodes during **electrolysis**. Reduction occurs at the cathode and oxidation at the anode.

Example During the electrolysis of molten lead (II) bromide:

cathode (–): $Pb^{2+} + 2e^{-} \longrightarrow Pb$ (reduction)

anode (+): $2Br^{-} - 2e^{-} \longrightarrow Br_2$ (oxidation)

This can also be written as $2Br^{-} \longrightarrow Br_2 + 2e^{-}$

Oxidation number

The **oxidation number**, or **oxidation state**, is the formal charge on an atom calculated on the basis that it is in a wholly ionic compound. Oxidation numbers are assigned according to several rules.

1. The oxidation numbers of the atoms in **uncombined elements** are zero.
2. For simple **monatomic ions**, the oxidation number is equal to the charge on the ion.
3. In a compound or **neutral molecule**, the sum of the oxidation numbers is zero.
4. In a **polyatomic ion**, the sum of the oxidation numbers is equal to the charge on the ion.
5. In any species, the more **electronegative atom** will be assigned a negative oxidation number and the less electronegative atom assigned a positive oxidation number.
6. Some elements are used as **reference points** when assigning oxidation number because they nearly always have invariable oxidation numbers in their compounds:

Na, K	+1
Mg, Ca	+2
Al	+3
H	+1 (except in metal hydrides)
F	–1
Cl	–1 (except in compounds with oxygen and fluorine)
O	–2 (except in peroxides, superoxides and fluorides)

Example What is the oxidation number of iodine in the following species? $\mathbf{I_2}$, $\mathbf{I^-}$, $Mg\mathbf{I}_2$, $\mathbf{I}Cl_3$, $\mathbf{I}O_3^-$

Answers

$\mathbf{I}_2 = 0$, as an uncombined element

$\mathbf{I}^- = -1$, as a simple monatomic ion

$Mg\mathbf{I}_2 = -1$, as Mg^{2+} has oxidation number +2, so $(+2) + (2 \times \mathbf{I}) = 0$, therefore each $\mathbf{I} = -1$

$\mathbf{I}Cl_3 = +3$, as Cl has oxidation number −1, so $(\mathbf{I}) + (3 \times -1) = 0$, therefore $\mathbf{I} = +3$

$\mathbf{I}O_3^- = +5$, as $(\mathbf{I}) + (3 \times -2) = -1$, therefore $\mathbf{I} = +5$

Naming

For compounds or ions containing elements that have a variable oxidation number, Roman numerals are used to indicate the oxidation number of the element concerned, and so name the chemical species. This is called **Stock notation**, after the chemist A. Stock who devised the method.

Examples

Compounds

copper (I) oxide	Cu_2O
copper (II) oxide	CuO
iron (II) sulphate	$FeSO_4$
phosphorus (V) chloride	PCl_5

Cations		**Anions**	
iron (II)	Fe^{2+}	chlorate (I)	ClO^-
tin (II)	Sn^{2+}	chlorate (V)	ClO_3^-
copper (I)	Cu^+	sulphate (VI)	SO_4^{2-}

Redox processes

When an element is oxidised, its oxidation number increases.
When an element is reduced, its oxidation number decreases.

Example

$2Na_{(s)} + Cl_{2(g)} \longrightarrow 2NaCl_{(s)}$

0 0 +1 −1

In this reaction, the oxidation number of Na has increased from 0 to +1; Na has been oxidised. The oxidation number of Cl has decreased from 0 to −1; Cl has been reduced.

In a redox reaction, **aX + bY → products**

If the oxidation number of X changes by +x and the oxidation number of Y changes by −y, then a(+x) + b(−y) = 0

Example

$2I^{-}_{(aq)} + 2Fe^{3+}_{(aq)} \longrightarrow I_{2(aq)} + 2Fe^{2+}_{(aq)}$

−1 +3 0 +2

For I, oxidation number increase = +1. For Fe, oxidation number decrease = −1. Applying the general formula, 2(+1) + 2(−1) = 0

Ionic half-equations

Equations for redox processes can be split into two **ionic half-equations**. Electrons pass from the reducing agent to the oxidising agent.

Example

$Zn_{(s)} + Cu^{2+}_{(aq)} \longrightarrow Zn^{2+}_{(aq)} + Cu_{(s)}$

In this example, zinc is the reducing agent (itself oxidised):
$Zn_{(s)} \longrightarrow Zn^{2+}_{(aq)} + 2e^{-}$

Copper (II) ions act as the oxidising agent (itself reduced):
$Cu^{2+}_{(aq)} + 2e^{-} \longrightarrow Cu_{(s)}$

The electrons lost by the zinc atoms are the same electrons as those gained by the copper (II) ions. Addition of the two ionic half-equations gives the overall redox reaction:

$Zn_{(s)} + Cu^{2+}_{(aq)} + 2e^- \longrightarrow Zn^{2+}_{(aq)} + Cu_{(s)} + 2e^-$

Written with the **e^- cancelled**: $Zn_{(s)} + Cu^{2+}_{(aq)} \longrightarrow Zn^{2+}_{(aq)} + Cu_{(s)}$

In general, the number of electrons in each half must be the same when ionic half-equations are combined. To do this, one or both half-reactions may have to be multiplied by an integer.

Example The reaction between iodide ions and iron (III) ions

Iodide ions are oxidised to iodine molecules:

$2I^-_{(aq)} \longrightarrow I_{2(aq)} + 2e^-$ ... (i)

Iron (III) ions are reduced to iron (II) ions:

$Fe^{3+}_{(aq)} + e^- \longrightarrow Fe^{2+}_{(aq)}$ ... (ii)

Equation (ii) is multiplied by 2 and then added to equation (i):

$2I^-_{(aq)} + 2Fe^{3+}_{(aq)} + 2e^- \longrightarrow I_{2(aq)} + 2Fe^{2+}_{(aq)} + 2e^-$

so allowing the e^- to be cancelled:

$2I^-_{(aq)} + 2Fe^{3+}_{(aq)} \longrightarrow I_{2(aq)} + 2Fe^{2+}_{(aq)}$

You may come across more complicated half-equations, but the basic principles remain the same.

Example Potassium dichromate (VI) is an important oxidising agent that only works in an acidic medium. It is reduced to chromium (III) ions, whilst the $H^+_{(aq)}$ ions from the acid end up as water. Each Cr atom undergoes a 3-electron reduction, but every dichromate ion contains 2 Cr atoms. Therefore 6 electrons appear on the left-hand side of the half-equation:

$Cr_2O_7^{2-}{}_{(aq)} + 14H^+_{(aq)} + 6e^- \longrightarrow 2Cr^{3+}_{(aq)} + 7H_2O_{(l)}$

If the above half-equation is combined with that for the oxidation of iron (II) to iron (III) ions, $Fe^{2+}_{(aq)} \longrightarrow Fe^{3+}_{(aq)} + e^-$, then, before combining the two half-equations, the second is multiplied by 6, giving an overall equation for the redox reaction as:

$$Cr_2O_7^{2-}{}_{(aq)} + 14H^+{}_{(aq)} + 6Fe^{2+}{}_{(aq)} \longrightarrow 6Fe^{3+}{}_{(aq)} + 7H_2O_{(l)} + 2Cr^{3+}{}_{(aq)}$$

Disproportionation is when a single species undergoes both oxidation and reduction in the same reaction.

Example

$$Cl_{2(aq)} + 2OH^-{}_{(aq)} \longrightarrow ClO^-{}_{(aq)} + Cl^-{}_{(aq)} + H_2O_{(l)}$$

In chlorine molecules, Cl_2, the oxidation number of Cl = 0

In chlorate (I) ions, ClO^-, the oxidation number of Cl = +1 (chlorine has been oxidised from 0 to +1)

In chloride ions, Cl^-, the oxidation number of Cl = −1 (chlorine has been reduced from 0 to −1)

Therefore a single species (Cl_2) has been simultaneously both oxidised and reduced. For this to be possible, the species concerned must be in an **intermediate** oxidation state.

Check yourself

1 Calculate the oxidation numbers of:

(a) N in N_2
(b) Ca in Ca
(c) N in N_2O_4
(d) N in NO_3^-
(e) Mn in MnO_4^-
(f) Mn in Mn_2O_3
(g) O in H_2O_2
(h) S in SO_3^{2-}
(i) O in F_2O
(j) S in $S_4O_6^{2-}$ (10)

2 Name the following compounds using Roman numerals to indicate the oxidation number of the metal:

(a) CrF_3
(b) $Fe(OH)_3$
(c) FeS
(d) $MnCO_3$
(e) $CuSO_4 . 5H_2O$ (5)

3 Use oxidation numbers to work out which species have been oxidised and which have been reduced in the following reactions:

(a) $2I^-_{(aq)} + Cl_{2(aq)} \longrightarrow 2Cl^-_{(aq)} + I_{2(aq)}$ (2)
(b) $2Fe^{3+}_{(aq)} + Sn^{2+}_{(aq)} \longrightarrow 2Fe^{2+}_{(aq)} + Sn^{4+}_{(aq)}$ (2)
(c) $Ca_{(s)} + H_{2(g)} \longrightarrow CaH_{2(s)}$ (2)

4 For each of the reactions in Question 3, split up the equation into two ionic half-equations. (6)

5 Use the two half-equations given in each case to construct an overall equation for the reactions described in (a) to (c):

(a) hydrogen peroxide solution oxidising hydrogen sulphide in an acidic solution $H_2O_{2(aq)} + 2H^+_{(aq)} + 2e^- \longrightarrow 2H_2O_{(l)}$
$H_2S_{(g)} \longrightarrow 2H^+_{(aq)} + S_{(s)} + 2e^-$ (1)

(b) iodine solution oxidising an aqueous solution of sodium thiosulphate $I_{2(aq)} + 2e^- \longrightarrow 2I^-_{(aq)}$
$S_2O_3^{2-}{}_{(aq)} \longrightarrow \frac{1}{2}S_4O_6^{2-}{}_{(aq)} + e^-$ (1)

(c) an acidified solution of potassium manganate (VII) oxidising aqueous hydrogen peroxide solution
$MnO_4^-{}_{(aq)} + 8H^+_{(aq)} + 5e^- \longrightarrow Mn^{2+}_{(aq)} + 4H_2O_{(l)}$
$H_2O_{2(aq)} \longrightarrow O_{2(g)} + 2H^+_{(aq)} + 2e^-$ (1)

The answers are on page 113.

The elements in the **periodic table** are arranged in order of ascending atomic number. Horizontal rows of elements are called **periods**.

Examples Period 2: lithium to neon

Period 3: sodium to argon

Vertical columns of elements are called **groups**.

Examples Group 1: lithium to francium

Group 7: fluorine to astatine

Melting and boiling temperatures

The **melting temperatures** of the elements hydrogen to argon are shown in the graph.

The graph of **boiling temperature** against atomic number would show a similar pattern, but the corresponding numerical values would be higher than those for melting temperature.

The structure and bonding of the elements sodium to argon is shown in the table below.

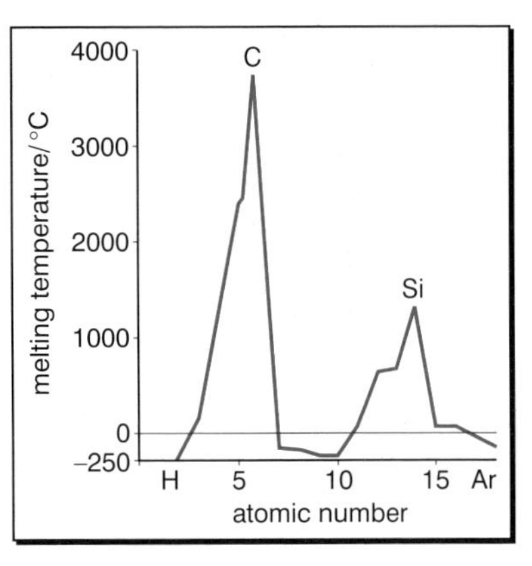

Element	Na Mg Al	Si	P S Cl	Ar
Bonding	⟵ Metallic ⟶	Covalent	⟵ Covalent ⟶	Single atoms
Structure	⟵ Giant metallic ⟶	Giant atomic (giant molecular)	⟵ Simple ⟶ molecules P_4 S_8 Cl_2	Ar

Main group chemistry

A study of **group 1** (lithium to caesium) and **group 2** (beryllium to barium) illustrates the behaviour of **metals** and their compounds. Conversely, a study of **group 7** (chlorine to iodine) illustrates the behaviour of a group of **non-metals** and their compounds.

Groups 1 and 2

The elements in group 1 comprise the **alkali metals**, whereas those in group 2 are the **alkaline-earth metals**.

Flame colours

The flame colours of the compounds of several group 1 and 2 metals are shown below.

Group 1	Group 2
Lithium: carmine red	Calcium: brick red
Sodium: yellow	Strontium: crimson red
Potassium: lilac	Barium: apple green

- A piece of nichrome wire is first cleaned by dipping it into concentrated hydrochloric acid.
- It is then dipped into a sample of the solid being tested.
- The wire is then held in a blue Bunsen flame and the colour of the flame is recorded.

Example Sodium chloride, NaCl

- The heat of the flame vaporises the compound, producing some sodium and chlorine atoms (electron configuration of Na: $1s^2\ 2s^2\ 2p^6\ 3s^1$).
- Heat energy from the flame promotes electrons from the 3s subshell to the 3p subshell in the sodium atoms.
- The electrons that were promoted fall back down to the 3s subshell, giving out energy in the form of light.

- The energy gap (ΔE) between the 3p and 3s subshells corresponds to the frequency of yellow light in the visible region of the spectrum.

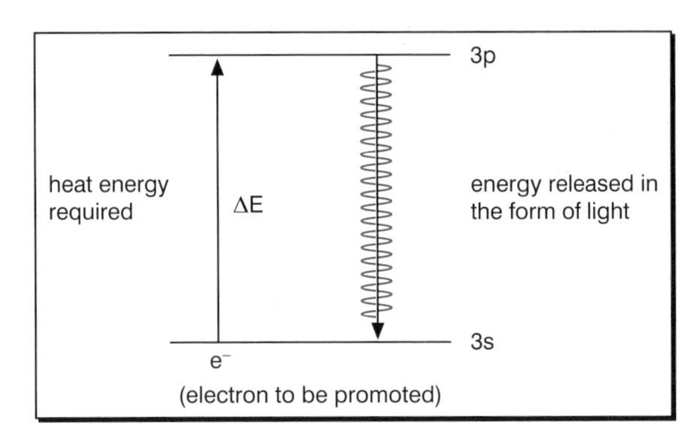

Chemical reactions

The elements become more reactive down groups 1 and 2. The atoms of the elements lower down the groups lose electrons more easily than those higher up, as illustrated by the trends in **ionisation energies** of the elements.

All group 1 metals **burn in oxygen**.

Example Lithium forms lithium oxide

$$4Li_{(s)} + O_{2(g)} \longrightarrow 2Li_2O_{(s)}$$

The oxide anion is O^{2-}.

Sodium produces a mixture of sodium oxide, Na_2O, and sodium peroxide, Na_2O_2.

$$2Na_{(s)} + O_{2(g)} \longrightarrow Na_2O_{2(s)}$$

sodium peroxide

The peroxide anion is O_2^{2-}.

The peroxide of sodium is the major product in excess oxygen.

The remaining alkali metals react with oxygen to form **superoxides**.

Examples Potassium and rubidium

$K_{(s)} + O_{2(g)} \longrightarrow KO_{2(s)}$

potassium superoxide

$Rb_{(s)} + O_{2(g)} \longrightarrow RbO_{2(s)}$

rubidium superoxide

The superoxide anion is unusual because it contains an unpaired electron.

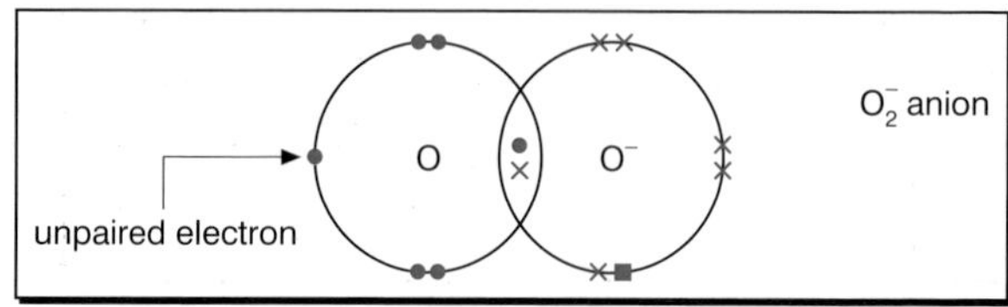

Group 1 superoxides are coloured compounds, which is unusual for group 1 compounds not containing a transition metal ion. The trend in oxide formation is the result of the increasing size of the metal cations as the group is descended.

The group 2 metals all burn to form **simple oxides** containing the O^{2-} ion.

Example $2Mg_{(s)} + O_{2(g)} \longrightarrow 2MgO_{(s)}$

The only exception is barium which, in excess oxygen, forms a mixture of barium oxide, BaO, and barium peroxide, BaO_2.

Group 1 elements all react **vigorously with cold water** to produce the metal hydroxide (an alkali) and hydrogen gas.

Example $2Na_{(s)} + 2H_2O_{(l)} \longrightarrow 2NaOH_{(aq)} + H_{2(g)}$

In group 2, beryllium does not react with water and magnesium reacts slowly with cold water. With steam, however, magnesium burns brightly to produce magnesium oxide and hydrogen.

$Mg_{(s)} + H_2O_{(g)} \longrightarrow MgO_{(s)} + H_{2(g)}$

Because the reactivity of the metals increases down the group, the remaining metals react vigorously with cold water.

Example $Ba_{(s)} + 2H_2O_{(l)} \longrightarrow Ba(OH)_{2(aq)} + H_{2(g)}$

The metal hydroxide is often produced as a suspension in water.

Solubility of group 2 hydroxides and sulphates

Solubility data for group 2 **hydroxides** in water at 298 K are given below.

Compound	Solubility/mol per 100 g of water
$Mg(OH)_2$	0.020×10^{-3}
$Ca(OH)_2$	1.5×10^{-3}
$Sr(OH)_2$	3.4×10^{-3}
$Ba(OH)_2$	15×10^{-3}

As the group is descended, the solubility of the group 2 hydroxides increases. $Mg(OH)_2$ is insoluble, both $Ca(OH)_2$ and $Sr(OH)_2$ are slightly soluble and $Ba(OH)_2$ is fairly soluble.

Solubility data for group 2 **sulphates** in water at 298 K are given below.

Compound	Solubility/mol per 100 g of water
$MgSO_4$	3600×10^{-4}
$CaSO_4$	11×10^{-4}
$SrSO_4$	0.71×10^{-4}
$BaSO_4$	0.0090×10^{-4}

As the group is descended, the solubility of the group 2 sulphates decreases. $MgSO_4$ is soluble, $CaSO_4$ is slightly soluble, whereas $SrSO_4$ and $BaSO_4$ are insoluble.

Thermal stability of nitrates and carbonates

Down groups 1 and 2, the **thermal stability** of the **nitrates** and **carbonates** increases.

Group 2 nitrates (and lithium nitrate) decompose to give the metal oxide, nitrogen dioxide (a brown gas) and oxygen.

Examples $2Ca(NO_3)_{2(s)} \longrightarrow 2CaO_{(s)} + 4NO_{2(g)} + O_{2(g)}$

$4LiNO_{3(s)} \longrightarrow 2Li_2O_{(s)} + 4NO_{2(g)} + O_{2(g)}$

Group 1 nitrates (except lithium nitrate) decompose to give the metal nitrite and oxygen.

Example $2NaNO_{3(s)} \longrightarrow 2NaNO_{2(s)} + O_{2(g)}$

Group 1 carbonates are stable to heat at the temperatures of a Bunsen flame, **except** lithium carbonate.

$Li_2CO_{3(s)} \longrightarrow Li_2O_{(s)} + CO_{2(g)}$

Group 2 carbonates (except barium carbonate which is stable to heat) all decompose to the metal oxide and carbon dioxide.

Example $CaCO_{3(s)} \longrightarrow CaO_{(s)} + CO_{2(g)}$

Thermal decomposition occurs more readily in compounds where the metal cation polarises the anion. Therefore:

- group 2 compounds (where the cation has a 2+ charge) are more likely to decompose than group 1 compounds (where the cation has a 1+ charge);
- compounds at the top of groups 1 and 2 (containing smaller cations) decompose more easily than those of elements lower down the groups;
- smaller cations have a higher charge density and are, therefore, more polarising.

Group 7

The elements in **group 7** comprise the **halogens**. Some information about the halogens chlorine to iodine is given in the table below.

Element	Symbol	Formula, colour and physical state at 20°C	Melting temperature/ °C	Boiling temperature/ °C	Atomic radius /nm	Ionic radius /nm
Chlorine	Cl	Cl_2 green gas	–101	–35	0.071	0.133
Bromine	Br	Br_2 red-brown liquid	–7	59	0.114	0.195
Iodine	I	I_2 dark grey shiny solid	114	184	0.132	0.215

Chlorine (Cl_2) turns damp blue litmus paper red, then bleaches it.

$$Cl_{2(g)} + H_2O_{(l)} \longrightarrow HCl_{(aq)} + \underset{\text{chloric (I) acid}}{HClO_{(aq)}}$$

Bromine (Br_2) displaces iodine from a solution of potassium iodide and, as a consequence, the iodine turns starch blue-black. (Starch–iodide paper may also be used.)

$$Br_{2(aq)} + 2I^-_{(aq)} \longrightarrow I_{2(aq)} + 2Br^-_{(aq)}$$

Iodine (I_2) turns starch solution from colourless to blue-black.

Hydrogen halides

Hydrogen halides are compounds formed on reaction of the halogen with hydrogen.

$$H_{2(g)} + X_{2(g)} \longrightarrow 2HX_{(g)}$$

The hydrogen halides are colourless gases at room temperature and pressure. They produce steamy fumes in moist air. They are covalently bonded molecules, with simple molecular structures. They are very soluble in water, as they react to form ions.

Example $$HCl_{(g)} \xrightarrow{+\,aq} H^+_{(aq)} + Cl^-_{(aq)}$$

The resulting solutions are strongly acidic.

Solutions of **halide ions** in **water** can be tested for:

- add a few drops of dilute nitric acid, to rule out the possibility of carbonate and/or sulphite ions being present;
- add aqueous silver nitrate solution, followed by ammonia solution.

Test	Chloride	Bromide	Iodide
Addition of aqueous silver nitrate	White precipitate	Cream precipitate	Yellow precipitate
Addition of dilute $NH_{3(aq)}$	Precipitate dissolves	No change	No change
Addition of concentrated $NH_{3(aq)}$	Precipitate dissolves	Precipitate dissolves	No change

The general equation for the formation of the silver halide precipitate is:

$Ag^{+}{}_{(aq)} + X^{-}{}_{(aq)} \longrightarrow AgX_{(s)}$

The reactions of the **halide salts**, in the solid state, with **concentrated sulphuric acid** vary in relation to the reducing power of the hydrogen halides produced.

Example $NaX_{(s)} + H_2SO_{4(l)} \longrightarrow NaHSO_{4(s)} + HX_{(g)}$

Because concentrated sulphuric acid is also an **oxidising agent**, hydrogen bromide and hydrogen iodide are oxidised to the free halogens bromine and iodine, respectively. The results are shown in the table below.

	Name of solid halide		
	Sodium chloride	**Sodium bromide**	**Sodium iodide**
Addition of concentrated sulphuric acid	Steamy fumes of $HCl_{(g)}$	Steamy fumes of $HBr_{(g)}$ with some red-brown vapour of bromine	Clouds of purple iodine vapour and black solid residue

Hydrogen iodide is the strongest **reducing agent**. This trend is expected when the bond enthalpies of the hydrogen halides are considered.

Bond	Bond enthalpy/kJ mol^{-1}
H–Cl	+432
H–Br	+366
H–I	+298

As group 7 is descended, the **oxidising power** of the halogens decreases.

chlorine	↓	decreasing
bromine		oxidising
iodine		strength

Chlorine, being the most powerful oxidising agent of the three halogens listed, is the most readily reduced.

$$Cl_{2(aq)} + 2e^{-} \longrightarrow 2Cl^{-}_{(aq)}$$

An application of the above order of reactivity is the use of chlorine in the extraction of bromine from sea water.

$$Cl_{2(aq)} + 2Br^{-}_{(aq)} \longrightarrow 2Cl^{-}_{(aq)} + Br_{2(aq)}$$

The chlorine molecules are reduced and the bromide ions are oxidised.

Disproportionation is when a species undergoes both oxidation and reduction in the same reaction. In the manufacture of bleach, sodium chlorate (I), a disproportionation reaction occurs.

$$\underset{0}{\underline{Cl}_{2(aq)}} + 2NaOH_{(aq)} \longrightarrow \underset{+1}{Na\underline{Cl}O_{(aq)}} + \underset{-1}{Na\underline{Cl}_{(aq)}} + H_2O_{(l)}$$

Chlorine, in $Cl_{2(aq)}$, at oxidation number 0 has been both oxidised (to +1 in NaClO) and reduced (to –1 in NaCl).

At higher temperatures (70°C), the chlorate (I) ion also disproportionates.

$$\underset{+1}{3\underline{Cl}O^{-}_{(aq)}} \longrightarrow \underset{-1}{2\underline{Cl}^{-}_{(aq)}} + \underset{+5}{\underline{Cl}O_3{}^{-}_{(aq)}}$$

Check yourself

1 For the period sodium to argon, explain, in terms of structure and bonding, the variation in the melting temperatures of the elements.

Element	Na	Mg	Al	Si	P	S	Cl	Ar
Melting temperature/°C	98	649	660	1410	44	119	–101	–189

(6)

2 Explain the decrease in atomic radii of the elements in period 3.

Element	Na	Mg	Al	Si	P	S	Cl	Ar
Atomic radius/nm	0.19	0.16	0.13	0.12	0.11	0.10	0.099	0.096

(3)

3 State the colour of the flame that results when the following compounds are subjected to flame tests.

(a) Potassium chloride. (1)

(b) Sodium bromide. (1)

(c) Lithium iodide. (1)

4 Explain briefly the origin of the colours in flame tests. (2)

5 **(a)** Write an equation, including state symbols, for the reaction between calcium and water. (2)

(b) State the trend in solubility of the hydroxides of the group 2 elements as the atomic mass of the metal increases. (1)

6 Compare the reaction when sodium is burned in a plentiful supply of oxygen with that of magnesium and oxygen. Account briefly for any differences. (4)

7 When a sample of magnesium nitrate is heated, it gives off a brown gas and a gas that relights a glowing splint. Write an equation for this reaction. (2)

The answers are on page 114.

Check yourself

8 Account for the trend in decomposition temperatures of the group 2 carbonates, in terms of the sizes and charges of the cations involved.

Carbonate	$BeCO_3$	$MgCO_3$	$CaCO_3$	$SrCO_3$	$BaCO_3$
Decomposition temperature/°C	100	540	900	1290	1360

(3)

9 State what would be seen when concentrated sulphuric acid is added to separate samples of the following compounds.

(a) Sodium chloride. (1)

(b) Sodium bromide. (1)

(c) Sodium iodide. (1)

(d) What do these observations tell you about the relative ease of oxidation of the hydrogen halides? (2)

10 Sea water contains aqueous bromide ions. Chlorine gas is used in the extraction of bromine from this source.

(a) Give the ionic equation for the reaction between chlorine gas and bromide ions, including state symbols. (2)

(b) Explain which of the elements, chlorine or bromine, is the stronger oxidising agent in terms of electron transfer. (2)

The answers are on page 115.

Enthalpy change, ΔH

The **enthalpy change** of a reaction, ΔH, is the heat energy change when the reaction is carried out at constant pressure. It is necessary to express these values under **standard conditions**. For enthalpy changes measured under standard conditions, the symbol $\Delta H^\ominus$ is used. **Thermodynamic standard conditions** are:

- a quoted **temperature** which is often 298 K (25°C);
- a **pressure** of 1 atmosphere (100 kPa);
- elements in their **most stable states**, e.g. carbon as graphite (rather than diamond);
- aqueous solutions at a **concentration** of 1 mol dm^{-3}.

Enthalpy changes are normally measured in units of **kJ mol^{-1}**.

Reactions

An **exothermic reaction** is one in which heat energy is transferred (given out) to the surroundings. The sign of ΔH for an exothermic reaction is negative. An enthalpy level diagram for an exothermic reactions is:

Reactants

Enthalpy ↑ ↓ $\Delta H = -ve$

Products

An **endothermic reaction** is one in which heat energy is absorbed (taken in) from the surroundings. The sign of ΔH for an endothermic reaction is positive.

An enthalpy level diagram for an endothermic reaction is:

Products

Enthalpy ↑ ↑ $\Delta H = +ve$

Reactants

Thermodynamic definitions

There are several important definitions that you need to know.

Standard enthalpy of formation, $\Delta H_f^{\ominus}$, is the enthalpy change when one mole of a compound is formed, under standard conditions, from its elements in their standard states.

Example The standard enthalpy of formation of ethene refers to the reaction: $2C_{(s,\ graphite)} + 2H_{2(g)} \longrightarrow C_2H_{4(g)}$

By definition, the standard enthalpy of formation of all elements in their standard states, under standard conditions, is **zero**.

Example For graphite the enthalpy change for the reaction below is zero: $C_{(s,\ graphite)} \longrightarrow C_{(s,\ graphite)}$

Standard enthalpy of combustion, $\Delta H_c^{\ominus}$, is the enthalpy change when one mole of a substance is completely burned in oxygen, under standard conditions.

Example The standard enthalpy of combustion for methane refers to the reaction: $CH_{4(g)} + 2O_{2(g)} \longrightarrow CO_{2(g)} + 2H_2O_{(l)}$

Standard enthalpy of neutralisation, $\Delta H_{neut}^{\ominus}$, is the enthalpy change, under standard conditions, when one mole of water is produced as a result of the reaction between an acid and an alkali.

Example The standard enthalpy of neutralisation of dilute aqueous sulphuric acid with dilute aqueous sodium hydroxide refers to the equation:

$$\frac{1}{2}H_2SO_{4(aq)} + NaOH_{(aq)} \longrightarrow \frac{1}{2}Na_2SO_{4(aq)} + H_2O_{(l)}$$

For the reaction between dilute hydrochloric acid and dilute sodium hydroxide, however, the relevant equation is:

$$HCl_{(aq)} + NaOH_{(aq)} \longrightarrow NaCl_{(aq)} + H_2O_{(l)}$$

Because strong acids and strong alkalis are almost completely ionised, the standard enthalpy of neutralisation is the enthalpy change for the reaction:

$H^+_{(aq)} + OH^-_{(aq)} \longrightarrow H_2O_{(l)}$

The value for this enthalpy change is approximately constant at $\Delta H^{\ominus}_{neut} = -57.3$ kJ mol^{-1}.

If a weak acid or weak alkali is used (or both are weak), the standard enthalpy of neutralisation is generally less exothermic. This is because heat energy is required to ionise a weak acid or base.

Hess's law

Hess' law states that 'the enthalpy change accompanying a chemical reaction is independent of the pathway between the initial and final states'.

If a reaction is broken down into several intermediate steps, and each step is assigned an individual enthalpy change, then the sum of the individual changes must equal the overall enthalpy change, provided the initial and final states are the same.

By Hess's law

$\Delta H_{reaction} = \Delta H_1 + \Delta H_2 + \Delta H_3$

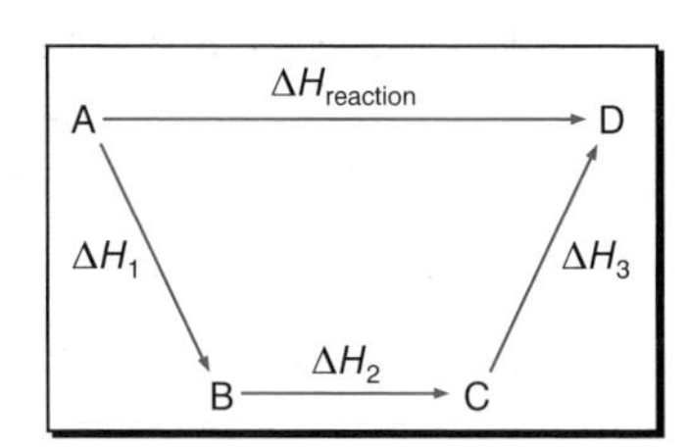

Calculating enthalpies

Enthalpies of reaction can be calculated from **enthalpy of formation data**.

Example Calculate the standard enthalpy change for the reaction:

$C_2H_{4(g)} + H_{2(g)} \longrightarrow C_2H_{6(g)}$

given the following standard enthalpies of formation:

$\Delta H^{\ominus}_f[C_2H_{6(g)}] = -85$ kJ mol^{-1} $\qquad$ $\Delta H^{\ominus}_f[C_2H_{4(g)}] = +52$ kJ mol^{-1}

Answer Construct a Hess's law cycle:

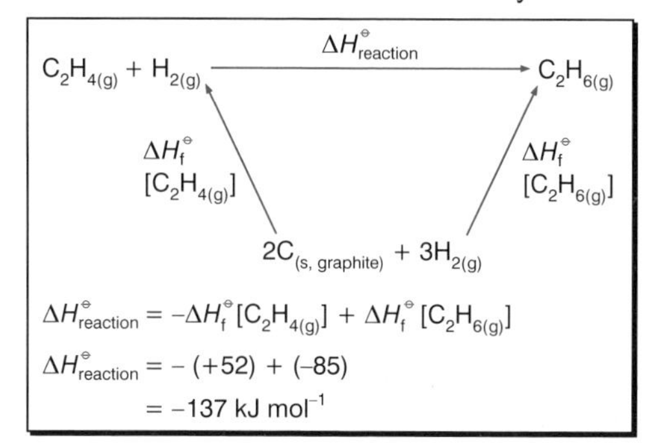

In general, for any reaction for which the enthalpies of formation of the products and reactants are known:

$\Delta H^\ominus_{reaction} = \Sigma\Delta H^\ominus_f$ of products $- \Sigma\Delta H^\ominus_f$ of reactants,

where the symbol Σ means 'the sum of'.

Remember that the enthalpies of formation of all elements are zero and that the enthalpies of formation must be scaled up accordingly when more than one mole of a compound appears in the equation.

Enthalpies of formation can be calculated from **combustion data**.

Example Calculate the standard enthalpy of formation of methane:

$C_{(s,\ graphite)} + 2H_{2(g)} \longrightarrow CH_{4(g)}$

given the following standard enthalpies of combustion:

Substance	$\Delta H^\ominus_c$ / kJ mol^{-1}
Carbon$_{(s,\ graphite)}$	–394
Hydrogen	–386
Methane	–891

Answer Construct a Hess's law cycle:

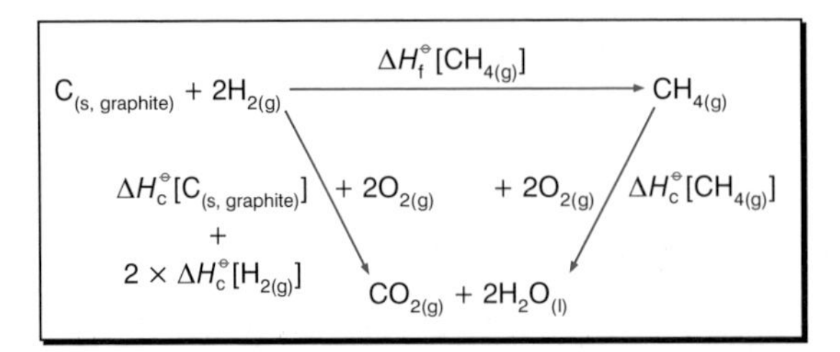

From the cycle it follows that:

$$\Delta H_f^\ominus\ [CH_{4(g)}] = \Delta H_c^\ominus\ [C_{(s,\ graphite)}] + 2 \times \Delta H_c^\ominus\ [H_{2(g)}] - \Delta H_c^\ominus\ [CH_{4(g)}]$$

$$\Delta H_f^\ominus\ [CH_{4(g)}] = (1 \times -394) + (2 \times -286) - (-891)$$

$$= -75\ \text{kJ mol}^{-1}$$

Enthalpies of reaction can also be calculated from **combustion data**.

Example Calculate the standard enthalpy change for the oxidation of ethanol (C_2H_5OH) to ethanal (CH_3CHO), given the following standard enthalpies of combustion:

$$\Delta H_c^\ominus\ [C_2H_5OH_{(l)}] = -1371\ \text{kJ mol}^{-1}$$

$$\Delta H_c^\ominus\ [CH_3CHO_{(g)}] = -1167\ \text{kJ mol}^{-1}$$

The balanced equation for the reaction is:

$$C_2H_5OH_{(l)} + \tfrac{1}{2}O_{2(g)} \longrightarrow CH_3CHO_{(g)} + H_2O_{(l)}$$

Answer Construct a Hess's law cycle:

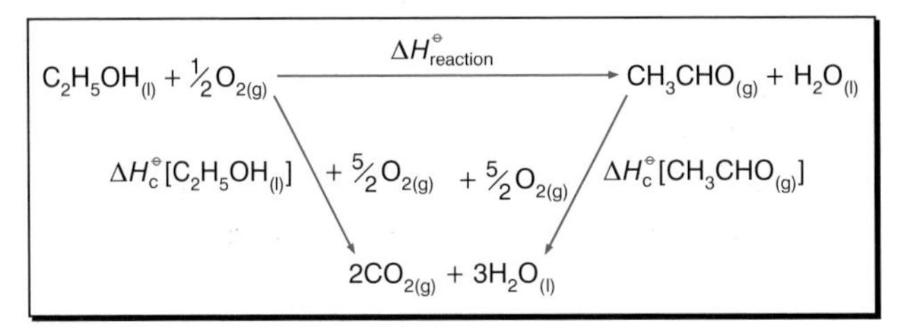

From the cycle it follows that:

$$\Delta H^{\ominus}_{reaction} = \Delta H^{\ominus}_{c}\,[C_2H_5OH_{(l)}] - \Delta H^{\ominus}_{c}\,[CH_3CHO_{(g)}]$$
$$= (-1371) - (-1167)$$
$$= -204\ \mathrm{kJ\,mol^{-1}}$$

Enthalpy changes of atomisation

The **standard enthalpy change of atomisation of an element, $\Delta H^{\ominus}_{at}$,** is the enthalpy change when one mole of gaseous atoms is produced from the element in its standard state, under standard conditions.

Examples $Na_{(s)} \longrightarrow Na_{(g)}$

$\frac{1}{2}Cl_{2(g)} \longrightarrow Cl_{(g)}$

In the case of solids, the enthalpy of atomisation includes the enthalpies of fusion and vaporisation; in the case of liquids, it includes the enthalpy of vaporisation.

The **standard enthalpy change of atomisation of a compound, $\Delta H^{\ominus}_{at}$,** is the enthalpy change when one mole of a compound is converted into gaseous atoms, under standard conditions.

Example $CO_{2(g)} \longrightarrow C_{(g)} + 2O_{(g)}$

Compare this definition with that for the standard enthalpy of atomisation of an element.

Alternatively, chemists use the expression **standard enthalpy change of dissociation**, which is the enthalpy change when one mole of a gaseous substance is broken up into free gaseous atoms, under standard conditions. This idea is used in the next section.

Average bond enthalpies

The energy required to break a particular bond depends on the nature of the two bonded atoms and the environment of the atoms.

Example In methane (CH_4) the four covalent carbon–hydrogen bonds are of different strength:

	$\Delta H^{\ominus}$/kJ mol^{-1}
$CH_{4(g)} \longrightarrow CH_{3(g)} + H_{(g)}$	+423
$CH_{3(g)} \longrightarrow CH_{2(g)} + H_{(g)}$	+480
$CH_{2(g)} \longrightarrow CH_{(g)} + H_{(g)}$	+425
$CH_{(g)} \longrightarrow C_{(g)} + H_{(g)}$	+335

The values +423, +480, +425 and +335 kJ mol^{-1} are referred to as the first, second, third and fourth bond dissociation enthalpies of methane, respectively. The **bond dissociation enthalpy** is the enthalpy required to break one mole of specific bonds in a specific molecule.

To calculate the **average bond enthalpy** of the (C–H) bond in methane, **$\overline{E}$ (C–H)**, the average of the four bond dissociation enthalpies is calculated:

$$(C\text{–}H) = \frac{+423 + 480 + 425 + 335}{4} = \frac{1663}{4} = +416\,\text{kJ mol}^{-1}$$

This is the enthalpy change for the reaction $\frac{1}{4}CH_{4(g)} \longrightarrow \frac{1}{4}C_{(g)} + H_{(g)}$.

Because average bond enthalpies refer to the endothermic process of bond breaking, they always have a positive sign.

In a diatomic molecule, e.g. chlorine (Cl_2), the bond dissociation enthalpy and the average bond enthalpy will have the same value. This is because both enthalpy changes refer to the process $Cl_{2(g)} \longrightarrow 2Cl_{(g)}$.

Because average bond enthalpies are average values calculated, for a particular bond, from a large number of compounds, they can only be used to determine **approximate enthalpy changes of reaction**. Average bond enthalpies are useful when estimating the values of enthalpy changes that cannot be determined directly. When calculating the enthalpy change of a reaction using average bond enthalpy values, it is important to remember:

- to use structural formulae of substances, if necessary, to work out which bonds have been broken and which have been made during a reaction;
- breaking bonds requires energy (endothermic process);
- making bonds releases energy (exothermic process);
- to calculate the overall enthalpy change, add the values for bond breaking (positive in sign) to those for bond making (negative in sign).

Example Use the average bond enthalpy data below:

$\overline{E}$ (C–C) = +348 kJ mol^{-1} $\overline{E}$ (C=C) = +612 kJ mol^{-1}

$\overline{E}$ (Cl–Cl) = +242 kJ mol^{-1} $\overline{E}$ (C–Cl) = +338 kJ mol^{-1}

to estimate the enthalpy change for the addition reaction between ethane and chlorine:

$C_2H_4 + Cl_2 \longrightarrow C_2H_4Cl_2$

Answer Use structural formulae to write the equation:

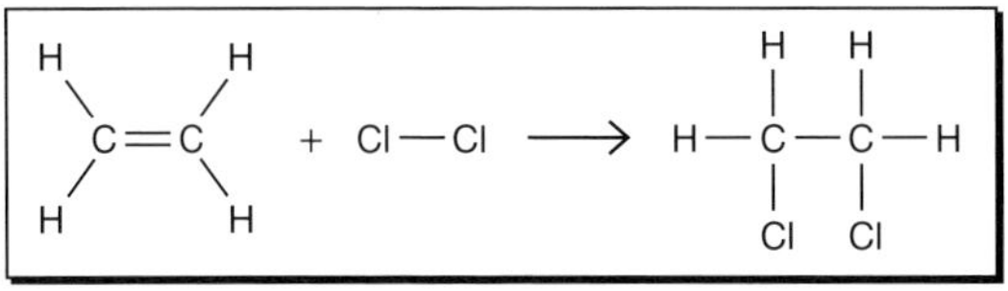

```
H         H                           H   H
 \       /                            |   |
  C === C     +  Cl—Cl   ——>     H—C—C—H
 /       \                            |   |
H         H                          Cl  Cl
```

Bonds broken		Bonds made	
$\overline{E}$ (C=C)	= +612 kJ mol^{-1}	– $\overline{E}$ (C–C)	= –348 kJ mol^{-1}
$\overline{E}$ (Cl–Cl)	= +242 kJ mol^{-1}	2 × – $\overline{E}$ (C–Cl)	= –676 kJ mol^{-1}
Total	= +854 kJ mol^{-1}	Total	= –1024 kJ mol^{-1}

Add the values for bond breaking and bond making

$\Delta H^{\ominus}_{reaction} = +854 + (-1024) = -170 \text{ kJ mol}^{-1}$

Check yourself

1 Draw an enthalpy level diagram for:

(a) the exothermic reaction $2H_{2(g)} + O_{2(g)} \longrightarrow 2H_2O_{(l)}$; $\Delta H = -572\,kJ\,mol^{-1}$. (1)

(b) the endothermic reaction $2HCl_{(g)} \longrightarrow H_{2(g)} + Cl_{2(g)}$; $\Delta H = +184\,kJ\,mol^{-1}$. (1)

2 State Hess's law. (2)

3 **(a)** Define 'standard enthalpy formation', $\Delta H_f^{\ominus}$. (3)

(b) Calculate the standard enthalpy change for the reaction:

$NH_{3(g)} + HCl_{(g)} \longrightarrow NH_4Cl_{(s)}$

given the following standard enthalpies of formation (in kJ mol^{-1}): $\Delta H_f^{\ominus}$ $[NH_{3(g)}] = -46$

$\Delta H_f^{\ominus}$ $[HCl_{(g)}] = -92$

$\Delta H_f^{\ominus}$ $[NH_4Cl_{(s)}] = -314$ (2)

4 **(a)** Define 'standard enthalpy of combustion', $\Delta H_c^{\ominus}$. (3)

(b) Calculate the standard enthalpy of formation of pentane, $C_5H_{12(l)}$, given the following standard enthalpies of combustion (in kJ mol^{-1}): (3)

Substance	$\Delta H_c^{\ominus}$/kJ mol^{-1}
$C_{(s,\ graphite)}$	–395
$H_{2(g)}$	–286
$C_5H_{12(l)}$	–3520

5 Calculate the standard enthalpy change for the oxidation of methanol to methanal:

$CH_3OH_{(l)} + \frac{1}{2}O_{2(g)} \longrightarrow HCHO_{(g)} + H_2O_{(l)}$

(methanol) (methanal)

given the following standard enthalpies of combustion:

$\Delta H_c^{\ominus}$ $[CH_3OH_{(l)}] = -725\,kJ\,mol^{-1}$

$\Delta H_c^{\ominus}$ $[HCHO_{(g)}] = -561\,kJ\,mol^{-1}$ (2)

The answers are on page 115.

Check yourself

6 Use the following information to calculate the average bond enthalpy of the N–H bond in ammonia, $NH_{3(g)}$:
$\Delta H_f^\ominus [NH_{3(g)}] = -46.2\ kJ\ mol^{-1}$
$\Delta H_{at}^\ominus [N_{2(g)}] = +473\ kJ\ mol^{-1}$
$\Delta H_{at}^\ominus [H_{2(g)}] = +218\ kJ\ mol^{-1}$ (3)

7 Use the average bond enthalpy data below:
$\overline{E}$ (C≡C) = $+835\ kJ\ mol^{-1}$
$\overline{E}$ (C=C) = $+610\ kJ\ mol^{-1}$
$\overline{E}$ (C–C) = $+346\ kJ\ mol^{-1}$
$\overline{E}$ (C–H) = $+413\ kJ\ mol^{-1}$
$\overline{E}$ (H–H) = $+436\ kJ\ mol^{-1}$
to estimate the standard enthalpy change for the reaction (in $kJ\ mol^{-1}$):

$2H_{2(g)} + HC{\equiv}CH_{(g)} \longrightarrow CH_3CH_{3(g)}$ (3)

The answers are on page 117.

The factors that control the **rate** of a chemical reaction are:

- concentration of a solution;
- pressure of gaseous reactants;
- temperature;
- surface area of solids;
- light energy, for photochemical reactions;
- catalysis.

Changes in the rate of a reaction can be explained by the **collision theory**. The three main assumptions of this are:

- reactant particles must **collide** if they are to react and form products;
- a collision is only successful if a minimum energy barrier, called the **activation energy**, is exceeded;
- a collision will not be successful unless the colliding particles are correctly **orientated** to one another.

Example $CO_{(g)} + NO_{2(g)} \longrightarrow CO_{2(g)} + NO_{(g)}$

The correct orientation is:

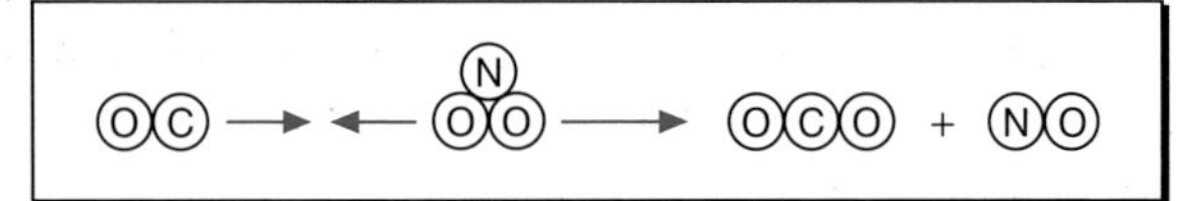

The incorrect orientation is:

The graph shows the energy profile for an exothermic reaction.

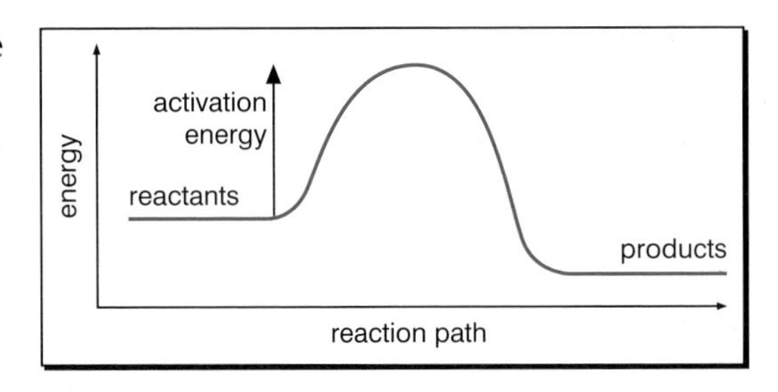

The activation energy is the **minimum kinetic energy** that the reactant particles must posses before they can collide to form products, i.e. collide 'successfully'.

Reactions with a large activation energy have a slow rate of reaction.

Concentration and pressure

An increase in the concentration of a **reactant** (or reactants) in solution, or an increase in the pressure on a gas-phase reaction, increases the rate of reaction. In terms of the collision theory:

- there are more reacting particles per unit volume, as the concentration increases;
- there is an increase in the frequency of collisions;
- the chance of a collision with energy greater than the activation energy also increases, as a consequence of the first two points.

Temperature

An increase in temperature increases the rate of a chemical reaction, because:

- the reacting particles move faster and so possess greater average kinetic energy;

therefore

- the particles collide more frequently;

but, more importantly,

- there is a greater fraction of collisions between particles with energy greater than the activation energy for the reaction.

Typically, for a 10 K rise in temperature, the rate of a chemical reaction doubles.

It can be shown that the number of particles that possess energy greater that the activation energy for a chemical reaction approximately doubles for a 10 K rise in temperature.

A study of the **Maxwell–Boltzmann** distribution of molecular energies at temperatures **T Kelvins** and at **(T + 10) K** illustrates this.

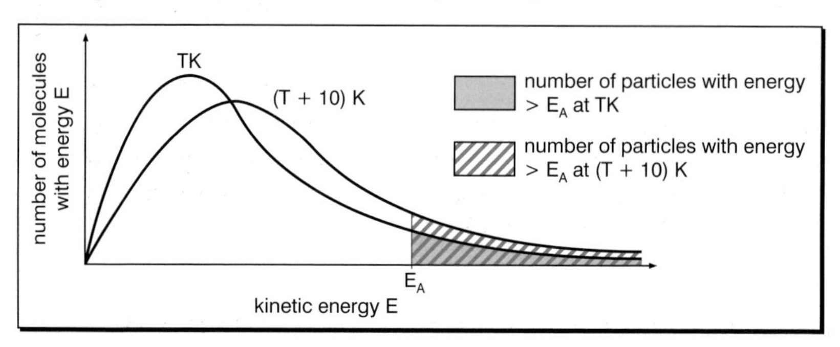

The area under the curve represents the total number of particles in the sample, and is constant.

The area under the curve beyond E_A (the activation energy) represents the number of particles that possess energy greater than the activation energy.

Surface area

As the state of **subdivision** of a solid increases, the surface area of the solid available to react becomes greater. Because reactions take place at the surface of a solid, the rate of reaction increases. This explains why solid chemicals are often supplied in powder form rather than in large lumps.

Examples

The rate of reaction between dilute hydrochloric acid and marble is increased when smaller pieces of marble are used.

$CaCO_{3(s)} + 2HCl_{(aq)} \longrightarrow CaCl_{2(aq)} + H_2O_{(l)} + CO_{2(g)}$

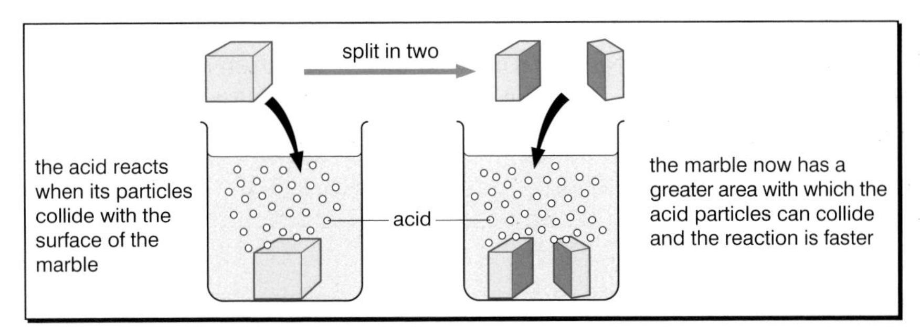

Explosions can occur in coal mines when coal dust, with a very large surface area, reacts with oxygen in the air without heating. Lumps of coal, with a smaller surface area than coal dust, do not react with oxygen unless heat energy is supplied.

$C_{(s)} + O_{2(g)} \longrightarrow CO_{2(g)}$

Similarly, in a mill, flour dust can explode in the presence of air.

Light

Many reactions are initiated by light. These are called **photochemical reactions**.

Examples

The reaction between hydrogen and chlorine.

$H_{2(g)} + Cl_{2(g)} \longrightarrow 2HCl_{(g)}$

The reaction occurs very slowly in the dark but it is explosive in sunlight.

Another example is photosynthesis.

$6CO_{2(g)} + 6H_2O_{(l)} \longrightarrow C_6H_{12}O_{6(aq)} + 6O_{2(g)}$

Light is a form of energy and can be used to initiate the breaking of bonds in reactant molecules.

Chemicals, such as hydrogen peroxide solution, are often stored in brown bottles in order to keep out light. This helps to slow down the decomposition.

$2H_2O_{2(aq)} \longrightarrow 2H_2O_{(l)} + O_{2(g)}$

Catalysts

Catalysts increase the rate of a reaction, but remain **chemically unchanged** at the end of the reaction. Catalysts provide an alternative reaction route of lower activation energy.

The **Maxwell–Boltzmann** distribution of molecular energies can be used to explain how a catalyst works at constant temperature.

At any particular temperature, more molecules will possess sufficient energy to overcome the activation energy for the catalysed reaction compared with the uncatalysed reaction. The distribution of molecular energies of the gas molecules does not change at constant temperature.

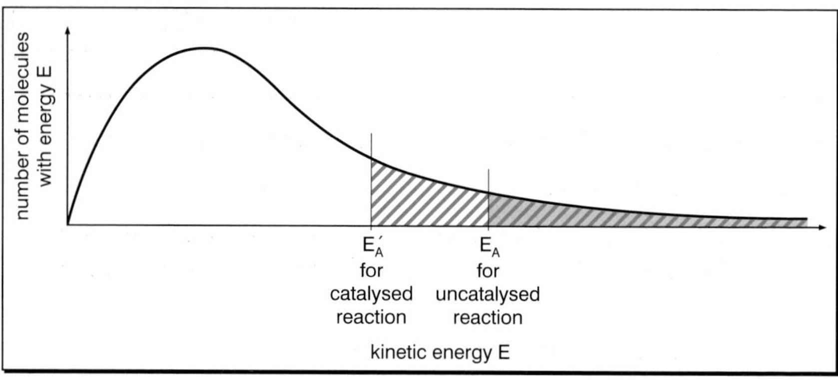

Thus the reaction proceeds **faster** when **catalysed**.

Example Iodide / peroxodisulphate (VI) reaction

Uncatalysed: $S_2O_8{}^{2-}{}_{(aq)} + 2I^-{}_{(aq)} \longrightarrow 2SO_4{}^{2-}{}_{(aq)} + I_{2(aq)}$

The reaction is catalysed by Fe^{2+}:

$$S_2O_8{}^{2-}{}_{(aq)} + 2Fe^{2+}{}_{(aq)} \longrightarrow 2SO_4{}^{2-}{}_{(aq)} + 2Fe^{3+}{}_{(aq)}$$

Then: $2Fe^{3+}{}_{(aq)} + 2I^-{}_{(aq)} \longrightarrow 2Fe^{2+}{}_{(aq)} + I_{2(aq)}$

↓ regenerated

Reaction profiles

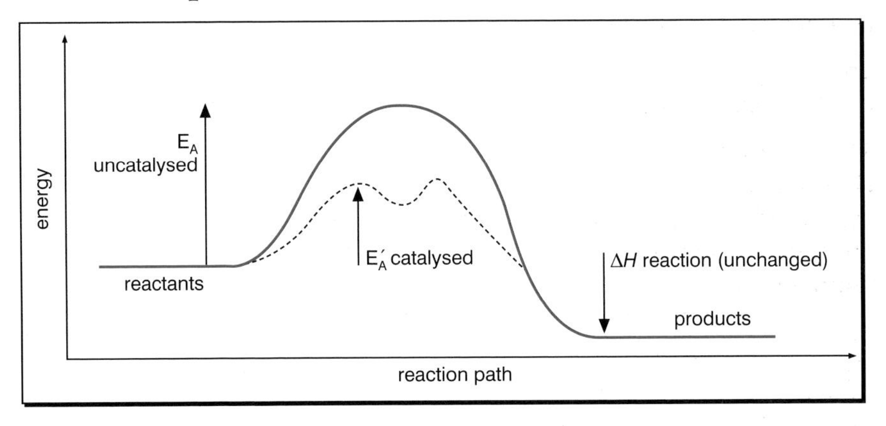

There are two types of catalysis: **homogeneous** and **heterogeneous**.

A **homogeneous catalyst** is in the same phase as the reactants.

Example The addition of a few drops of concentrated sulphuric acid catalyses the reaction between an alcohol and a carboxylic acid to form an ester

$$CH_3CO_2H_{(l)} + C_2H_5OH_{(l)} \rightleftharpoons CH_3CO_2C_2H_{5(l)} + H_2O_{(l)}$$

A **heterogeneous catalyst** is not in the same phase as the reactants.

Example Solid manganese (IV) oxide (MnO_2) increases the rate of decomposition of hydrogen peroxide, which is a liquid

$$2H_2O_{2(aq)} \longrightarrow 2H_2O_{(l)} + O_{2(g)}$$

Another example is the use of **catalytic metals** (platinum and rhodium) in the **catalytic converter** of a motor car. These solid metals catalyse the reaction between the pollutant gases carbon monoxide and nitrogen monoxide.

$2NO_{(g)} + 2CO_{(g)} \longrightarrow N_{2(g)} + 2CO_{2(g)}$

Cars with catalytic converters use lead-free petrol in order to prevent a lead coating forming on the surface of the catalyst.

Measuring the rate of a reaction

Reactions in which a gaseous product is formed are suitable to investigate in **rate of reaction** experiments.

Examples

acid + a carbonate ⟶ a salt + water + carbon dioxide

acid + a metal ⟶ a salt + hydrogen

The volume of gas evolved is measured at regular intervals and a graph of volume of gas given off versus time is plotted.

Example

Hydrochloric acid + magnesium ⟶ magnesium chloride + hydrogen

$2HCl_{(aq)} + Mg_{(s)} \longrightarrow MgCl_{2(aq)} + H_2(g)$

A suitable apparatus is shown below.

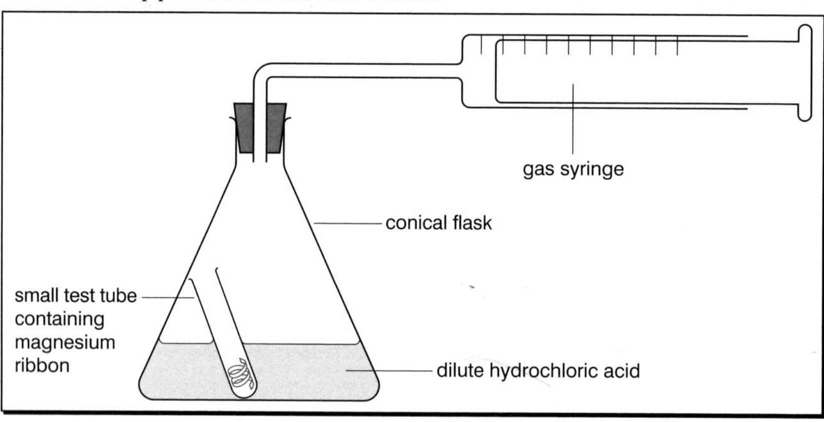

The magnesium is placed in a small test tube instead of dropping it directly into the acid and allowing gas to escape before the bung is secured in the flask. On shaking the flask, the acid is allowed to come into contact with the magnesium ribbon and the reaction commences, releasing hydrogen gas. The volume of gas produced is recorded at regular time intervals. A typical plot of the volume of hydrogen gas versus time is shown.

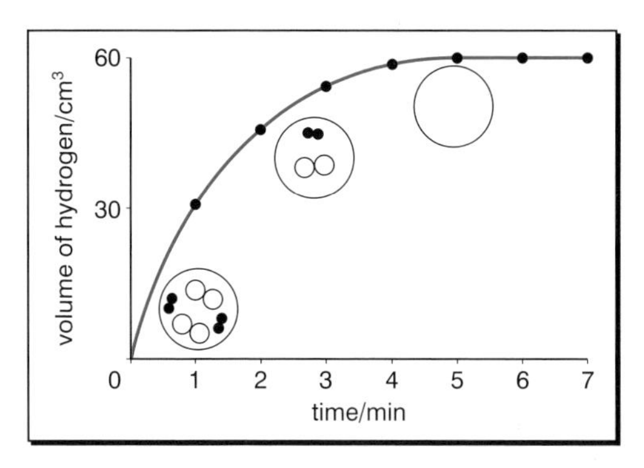

- The highest acid concentration ($H^+_{(aq)}$) / largest surface area of magnesium produces the steepest slope and fastest rate.
- As the concentration of acid / surface area of magnesium reduces, the slope of the graph lessens and the rate becomes slower.
- When the concentration of the acid is zero / the magnesium is all used up, the slope of the graph is zero and the reaction has finished.

Check yourself

1. List four factors that may control the rate of a chemical reaction. (4)
2. State the three major assumptions of collision theory. (3)
3. Define the term 'activation energy'. (2)
4. Use collision theory to explain why increasing the concentration of a reactant can increase the rate of a reaction. (3)
5. Use the Maxwell–Boltzmann graph of molecular energies to explain why increasing temperature increases the rate of a reaction. (4)
6. Define the term 'catalyst'. (2)
7. Explain briefly how the presence of a catalyst increases the rate of a reaction. (2)

The answers are on page 118.

Reversible reactions

Some chemical reactions are **irreversible** and proceed 100% to completion.

Example Between sodium metal and water

$2Na_{(s)} + 2H_2O_{(l)} \longrightarrow 2NaOH_{(aq)} + H_{2(g)}$

In such a reaction, a single arrow is drawn between the **reactants** (on the left-hand side) and the **products** (on the right-hand side).

Frying an egg is another familiar example of a 'one-way' chemical reaction. As the egg is cooked, chemical changes, involving the denaturation of proteins, occur. These processes cannot be reversed.

Other processes, including both chemical reactions and physical changes, can be **reversed**.

Examples

$$CuSO_4 . 5H_2O_{(s)} \underset{\text{add water}}{\overset{\text{heat}}{\rightleftarrows}} CuSO_{4(s)} + 5H_2O_{(l)}$$

hydrated copper (II) sulphate blue		anhydrous copper (II) sulphate white

$$CoCl_2 . 6H_2O_{(s)} \underset{\text{add water}}{\overset{\text{heat}}{\rightleftarrows}} CoCl_{2(s)} + 6H_2O_{(l)}$$

hydrated cobalt (II) chloride pink		anhydrous cobalt (II) chloride blue

The two reactions above are used to test, chemically, for the presence of water.

Water turning into steam is a well-known reversible physical change.

$$H_2O_{(l)} \underset{\text{cool}}{\overset{\text{heat}}{\rightleftarrows}} H_2O_{(g)}$$

Reversible reactions are therefore defined as reactions that can be made to proceed in one direction or the other by changing the conditions.

Chemical equilibrium

For the reversible reaction $A + B \rightleftarrows C + D$, if neither the forward nor the backward reaction is complete, after a period of time the reaction **apparently** ceases and a **state of equilibrium** is attained. All four species (A, B, C and D) are present in the mixture and the symbol '$\rightleftharpoons$' is used to denote equilibrium.

$A + B \rightleftharpoons C + D$

Chemical equilibria are, in fact, **dynamic equilibria** rather than static equilibria.

In the equilibrium shown opposite, the see-saw is not moving and a balance point has been achieved.

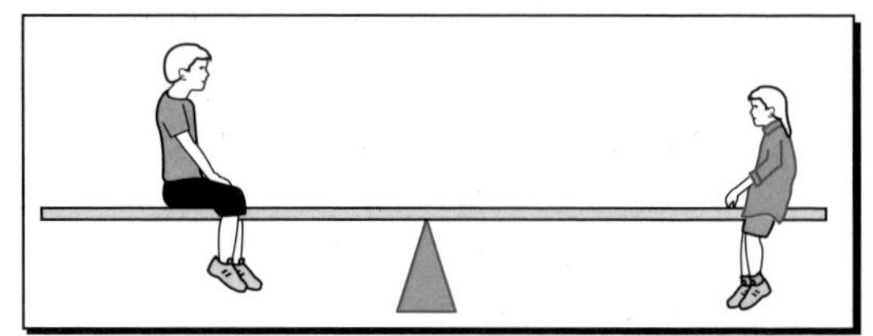

Chemical equilibrium is **not** like the see-saw above, but can be compared with the situation where a person is running up an escalator at the same rate as the escalator is descending. To the observer, the person stays in the same place on the escalator when dynamic equilibrium is attained.

In a chemical context, therefore, the system $A + B \rightleftharpoons C + D$ reaches a dynamic equilibrium when the rate of the forward reaction equals the rate of the backward reaction. At equilibrium, the concentrations of A, B, C and D remain constant.

Two further features of dynamic equilibria are:

- equilibrium can be achieved **from either direction**, provided that the temperature is the same, e.g. in the above system, an identical equilibrium is reached if starting with either A and B or C and D;
- equilibrium can only be attained in **a closed system**, i.e. none of the substances in the equilibrium should escape or other matter be able to enter the system.

The **iodine monochloride/iodine trichloride system** is an example of a chemical equilibrium. When chlorine is passed over iodine crystals, in a U-tube, a brown liquid is formed. This is iodine monochloride, $ICl_{(l)}$.

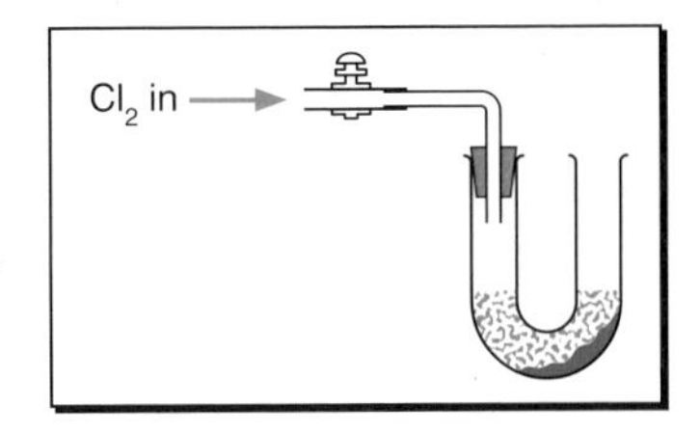

The equation is: $I_{2(s)} + Cl_{2(g)} \longrightarrow 2ICl_{(l)}$.

On addition of a little more chlorine gas, a yellow solid appears. This solid is iodine trichloride, $ICl_{3(s)}$. When the U-tube is stoppered, it contains a mixture of $ICl_{(l)}$, $Cl_{2(g)}$ and $ICl_{3(s)}$. Eventually, the number of moles of each component remains constant and a state of equilibrium attained.

$ICl_{(l)}$ +	$Cl_{2(g)}$	$\rightleftharpoons$	$ICl_{3(s)}$
brown liquid	green gas		yellow solid

Two graphs can be drawn to show what happens in the U-tube as equilibrium is achieved.

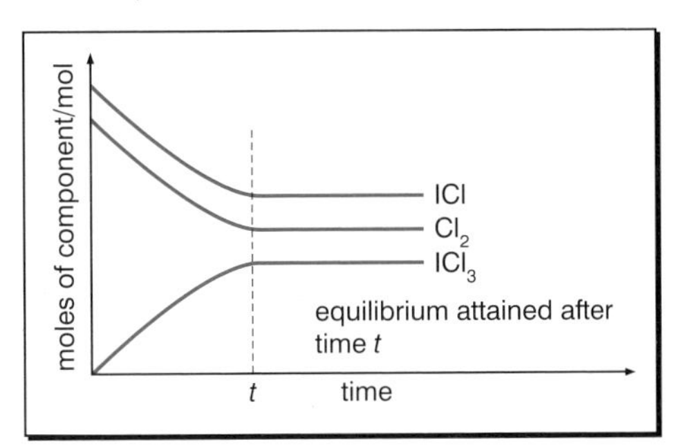

Alternatively, the rates of the forward reaction and the backward reaction can be plotted against time.

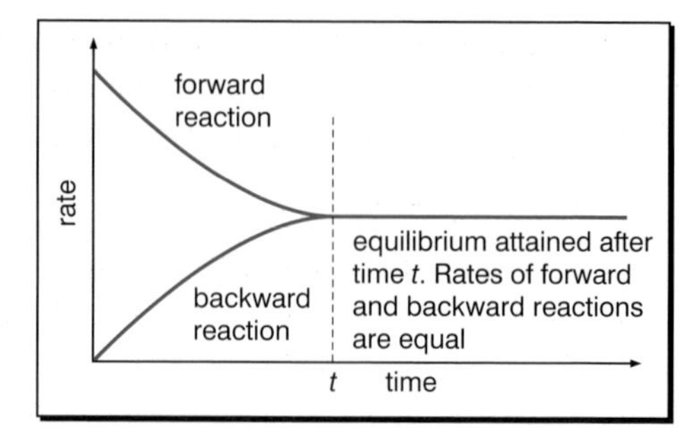

Furthermore, it can be observed that:

- when more chlorine is passed into the U-tube, more yellow solid forms and the brown liquid disappears;

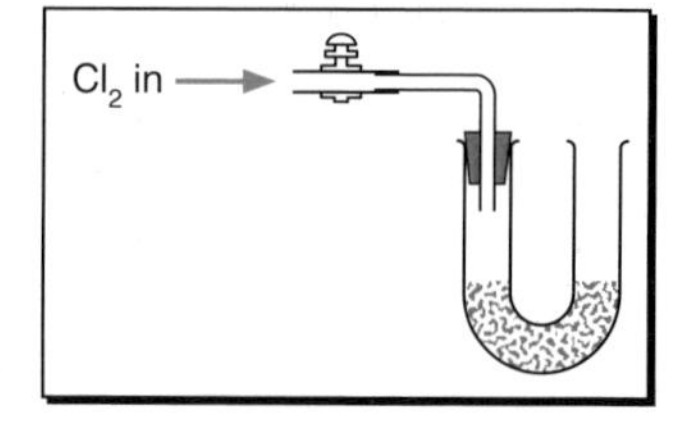

- when the U-tube is inverted, dense chlorine gas is removed from the system and more brown liquid reappears whilst the yellow solid is used up.

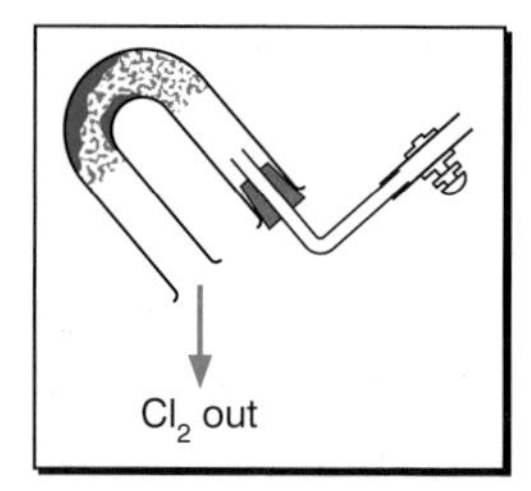

The **chromate (VI) / dichromate (VI) system** is another example of a chemical equilibrium. A few cm^3 of aqueous potassium chromate (VI) solution (containing yellow CrO_4^{2-} ions) are placed in a beaker. On addition of several drops of colourless dilute sulphuric acid, the yellow solution turns orange. The following equilibrium has been set up.

$$2CrO_4^{2-}{}_{(aq)} + 2H^+{}_{(aq)} \rightleftharpoons Cr_2O_7^{2-}{}_{(aq)} + H_2O_{(l)}$$

yellow   colourless   orange   colourless

The orange colour is due to the presence of dichromate (VI) ions, $Cr_2O_7^{2-}$. On addition of dilute sodium hydroxide solution, however, acidic $H^+_{(aq)}$ ions are removed by reaction with hydroxide ions.

$H^+_{(aq)} + OH^-_{(aq)} \longrightarrow H_2O_{(l)}$

The position of equilibrium shifts to the left-hand side and the yellow colour is restored. Addition of more dilute sulphuric acid increases the concentration of $H^+_{(aq)}$ ions, thereby shifting the position of equilibrium to the right-hand side and restoring the orange colour to the solution.

Factors affecting chemical equilibrium

Although it is **not** an explanation, **Le Chatelier's principle** is used to predict the effect of changes in conditions on the position of equilibrium. One statement of Le Chatelier's principle is: 'If a system in equilibrium is subjected to a change which disturbs the equilibrium, the system responds in such a way as to counteract the effect of the change'. The factors that may change the position of an equilibrium are concentration, temperature and pressure.

Concentration

The effect of concentration changes were observable in the two systems described above. For the system $A + B \rightleftharpoons C + D$, an increase in the concentration of A and/or B will shift the position of equilibrium to the right-hand side. For example, on increasing the concentration of A, some of the added A reacts with substance B to produce more C and D until equilibrium is re-established. Similarly, if the concentration of C and/or D is increased, the position of equilibrium is shifted to the left-hand side. Removal of a component, e.g. substance A, will cause the system to respond in such a way as to oppose the change, i.e. the decrease in the concentration of A. Therefore, the equilibrium position shifts to the left-hand side.

Temperature

If the temperature is **increased** (by adding heat energy), the position of equilibrium shifts to the **endothermic** direction. The system has responded in such a way as to remove the added heat energy. If the temperature is **lowered**, the position of equilibrium moves to the **exothermic** direction.

Example The Haber process used to manufacture ammonia

$N_{2(g)} + 3H_{2(g)} \rightleftharpoons 2NH_{3(g)}$ $\Delta H = -46\,\text{kJ mol}^{-1}$

Lowering the temperature shifts the position of equilibrium to the right, so increasing the yield of ammonia, NH_3.

Pressure

Changes in pressure only affect equilibria that involve gases and where there are different numbers of moles of gas molecules on the left- and right-hand side of the equilibrium. An increase in pressure shifts the position of equilibrium to the side with fewer moles of gas molecules.

Example $A_{(g)} + 3B_{(g)} \rightleftharpoons 2C_{(g)}$

left-hand side | right-hand side

4 moles of gas molecules $\rightleftharpoons$ 2 moles of gas molecules

Increasing the pressure shifts the position of the above equilibrium to the right. The system has responded to counteract the effect of the change (an increase in pressure). From kinetic theory, two moles of gas molecules exert less pressure on the walls of a vessel than four moles. As predicted, the position of equilibrium has altered in such a way as to reduce the total pressure.

Catalysts

A catalyst has no effect on the position of equilibrium. A catalyst increases the rate at which equilibrium is attained. As discussed in the Reaction Rates chapter, a catalyst provides an alternative route of lower activation energy. Because the rates of both the forward and backward reactions are increased, there is no change in the position of equilibrium. In industry, the presence of a catalyst allows a process to be carried out at a lower temperature (thereby reducing heat energy costs) whilst maintaining a viable rate of reaction.

Check yourself

1 Explain what is meant by the term 'dynamic equilibrium' as applied to the system $A + B \rightleftharpoons C + D$. (3)

2 List three factors that, when changed, may alter the position of a chemical equilibrium. (3)

3 Why does a catalyst increase the rate of a chemical reaction, yet have no effect on the position of equilibrium? (4)

4 State Le Chatelier's principle. (3)

5 Explain the effect (if any) of increasing the total pressure in the following equilibrium systems:

(a) $2SO_{2(g)} + O_{2(g)} \rightleftharpoons 2SO_{3(g)}$ (2)

(b) $H_{2(g)} + I_{2(g)} \rightleftharpoons 2HI_{(g)}$ (2)

6 Explain the effect (if any) of decreasing the temperature on the following equilibria:

(a) $2H_{2(g)} + CO_{(g)} \rightleftharpoons CH_3OH_{(g)}$ $\Delta H = -92\,kJ\,mol^{-1}$ (2)

(b) $AgClO_{2(s)} \rightleftharpoons Ag_{(s)} + \frac{1}{2}Cl_{2(g)} + O_{2(g)}$ $\Delta H = 0\,kJ\,mol^{-1}$ (2)

7 Predict the effect of adding dilute aqueous sodium hydroxide solution to each of the following equilibria:

(a) $Br_{2(aq)} + H_2O_{(l)} \rightleftharpoons 2H^{+}_{(aq)} + Br^{-}_{(aq)} + OBr^{-}_{(aq)}$ (2)

(b) $CH_3COOH_{(l)} + C_2H_5OH_{(l)} \rightleftharpoons CH_3COOC_2H_{5(l)} + H_2O_{(l)}$

ethanoic acid ethanol ethyl ethanoate (2)

The answers are on page 119.

The compounds of carbon are studied separately in **organic chemistry**. There are well over 2 million such compounds, because of the ability of carbon to form covalently bonded chains of carbon atoms. However, the study of organic chemistry does not include the chemistry of oxides of carbon or metal carbonates, e.g. calcium carbonate.

Compounds of carbon are grouped into a small number of 'families', called **homologous series** of compounds. In a homologous series, compounds have the **same general formula** and successive members of the series **differ by a – CH_2 – group**. The **chemical properties** of a homologous series are **similar** as each member contains the same functional group. However, there is a **gradation** (gradual change) in **physical properties** in an any particular series.

Alkanes

The **alkanes** make up a homologous series of **hydrocarbons** with the general formula C_nH_{2n+2} (*n* is the number of carbon atoms). Alkanes exhibit a type of **structural isomerism** called **chain isomerism**, where the arrangement of the carbon atoms in the molecules is different.

Example

```
    H   H   H   H   H   H
    |   |   |   |   |   |
H — C — C — C — C — C — C — H
    |   |   |   |   |   |
    H   H   H   H   H   H
          hexane
```

and

```
    H   H   H   H   H
    |   |   |   |   |
H — C — C — C — C — C — H
    |   |   |   |   |
    H   |   H   H   H
        |
    H — C — H
        |
        H
   2-methylpentane
```

Naming

Example The isomer 2-methylpentane

- The longest chain consists of 5 C atoms, so 'pentane'.
- A methyl group ($-CH_3$) is on the second C atom, so 2-methylpentane.

A hyphen is always put between a number and a letter.

Reactions

Alkanes burn in a plentiful supply (excess) of air or oxygen to form carbon dioxide and water. Because such **combustion reactions** are exothermic, alkanes are very useful as fuels.

Example Methane undergoes complete combustion

$$CH_{4(g)} + 2O_{2(g)} \longrightarrow CO_{2(g)} + 2H_2O_{(l)}$$

Alkanes react with halogens, such as chlorine and bromine, to form halogenoalkanes.

Example Chlorination of methane

$$CH_4 + Cl_2 \longrightarrow \underset{\text{chloromethane}}{CH_3Cl} + HCl$$

Conditions Ultraviolet (u.v.) light must be present

Chlorine molecules (Cl_2) absorb the u.v. light and the (Cl–Cl) covalent bonds break to form two chlorine atoms:

$$Cl_2 \longrightarrow Cl\cdot + Cl\cdot$$

This is a **free radical substitution reaction**. Because chlorine atoms have 7 outer shell electrons, each will possess an unpaired electron. So 2 chlorine **radicals** are produced. A radical is a species that has a **single unpaired electron**.

When a covalent bond breaks to produce radicals, i.e. one electron of the bond pair goes to each atom, **homolytic fission** has occurred. These highly reactive chlorine radicals attack the methane molecules.

Referring to the equation above, because a hydrogen atom has been replaced by a chlorine atom, a **substitution reaction** has taken place. A substitution reaction is when one atom (or group of atoms) in a molecule is **replaced** by **another** atom (or group of atoms).

The mechanism for the chlorination of methane is described in three stages:

Initiation: $Cl_2 \longrightarrow Cl\cdot + Cl\cdot$

Propagation:	$CH_4 + Cl\cdot$	$\longrightarrow$	$CH_3\cdot + HCl$
	$CH_3\cdot + Cl_2$	$\longrightarrow$	$CH_3Cl + Cl\cdot$
Termination:	$Cl\cdot + Cl\cdot$	$\longrightarrow$	Cl_2
	$CH_3\cdot + Cl\cdot$	$\longrightarrow$	CH_3Cl
	$CH_3\cdot + CH_3\cdot$	$\longrightarrow$	C_2H_6

Alkenes

The **alkenes** make up a homologous series of **hydrocarbons** with the general formula C_nH_{2n}. Alkenes show two types of structural isomerism, **position isomerism** and **chain isomerism**. **Geometrical isomerism** also exists because of the lack of free rotation about the C=C double bond.

Example

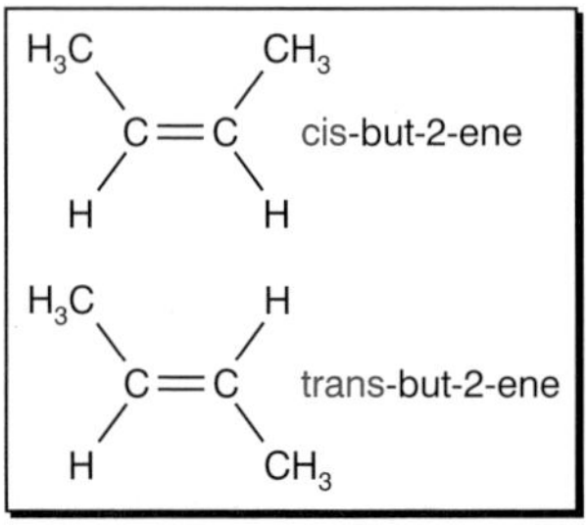

In the first isomer, identical groups are on the same side of the molecule; in the second, the groups are on opposite sides of the molecule. They take the names **cis** and **trans**, respectively. In general, geometrical isomerism occurs for:

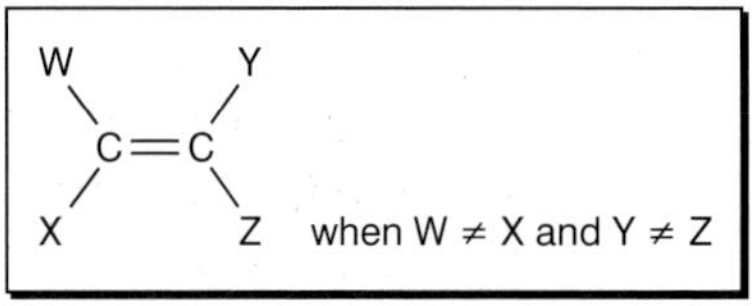

H_3C and CH_3 on the same side, H and H on the same side of C=C: cis-but-2-ene

H_3C and H above, H and CH_3 below, across C=C: trans-but-2-ene

W and Y above, X and Z below, across C=C: when W ≠ X and Y ≠ Z

Geometrical isomerism is a type of **stereoisomerism**: the compounds have the same molecular and structural formulae but the arrangement of their atoms in space is different.

Reactions

Alkenes are more reactive than alkanes because of the presence of the **pi (π) bond** (discussed on page 28). The π electrons in the C=C double bond represent the reactive site of the alkene. Because a π bond is weaker than a sigma (σ) bond, the π bond breaks to allow **addition reactions**. In an addition reaction, two molecules react together to form a single product.

Examples

Halogens: alkenes add bromine (or chlorine)

Conditions Bromine in an inert solvent at room temperature and pressure (r.t.p)

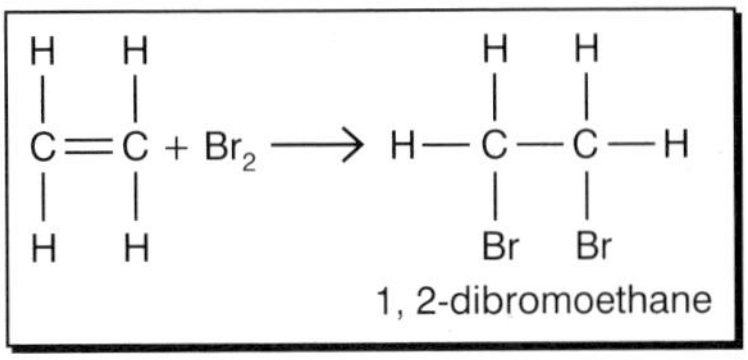

```
H   H                   H   H
|   |                   |   |
C = C  + Br2  -->  H -- C -- C -- H
|   |                   |   |
H   H                   Br  Br
                 1, 2-dibromoethane
```

Because the product is colourless, the decolorisation of a reddy-brown solution of bromine is used to test for the >C=C< group.

Hydrogen halides: alkenes add hydrogen halides across the double bond to form halogenoalkanes.

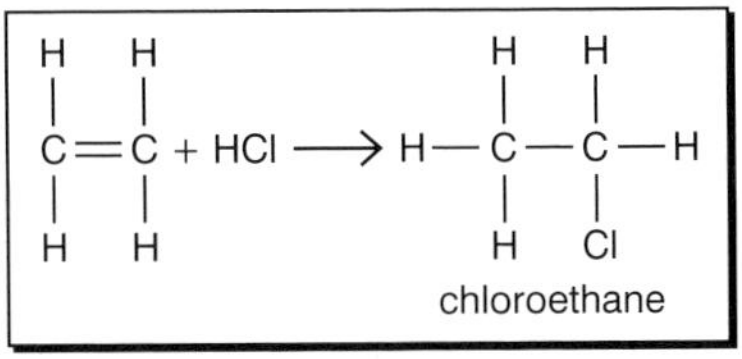

```
H   H                  H   H
|   |                  |   |
C = C  + HCl -->  H -- C -- C -- H
|   |                  |   |
H   H                  H   Cl
                  chloroethane
```

Conditions Add hydrogen chloride gas at r.t.p.

With unsymmetrical alkenes, **Markownikoff's rule** is applied. On addition of H–X, the hydrogen atom adds to the carbon atom which already has the more hydrogen atoms directly bonded to it. Applying the rule results in 2-bromopropane as the major product below.

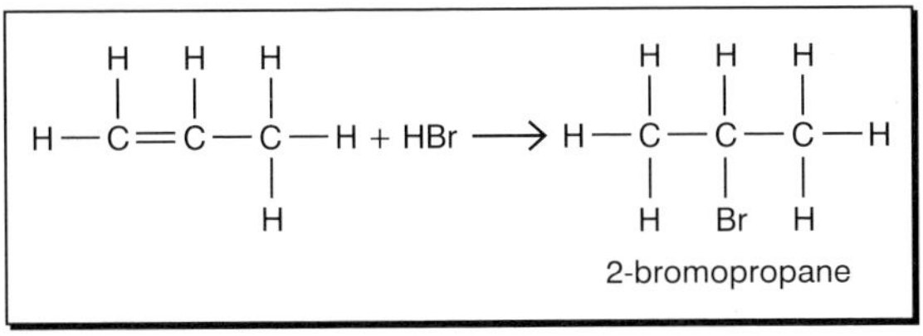

```
     H   H   H                      H   H   H
     |   |   |                      |   |   |
H -- C = C -- C -- H + HBr -->  H -- C -- C -- C -- H
             |                      |   |   |
             H                      H   Br  H
                                   2-bromopropane
```

Conditions Add hydrogen bromide gas at r.t.p.

The reactions of halogens and hydrogen halides with alkenes are **electrophilic addition reactions**. This means that the initial attack on the organic molecule is by an electron-deficient species that **accepts** a **lone pair** of electrons to form a covalent bond. This species is called an **electrophile**. In the case of the reaction with hydrogen bromide, the mechanism for the reaction is as shown.

The δ^+ hydrogen atom seeks out the electron density in the double bond. The curly arrow represents the movement of a pair of electrons.

The H–Br bond breaks, with both electrons in the bonding pair going to the Br atom. This is an example of **electrophilic fission** of a bond, forming H^+ and Br^-. The H^+ is added to the alkene and a carbocation intermediate is produced.

In the final step, a pair of electrons is donated by the Br^- ion to form the product molecule.

$$\mathrm{>C{=}C<} + H^{\delta+}{-}Br^{\delta-} \longrightarrow \mathrm{>\overset{+}{C}{-}C(H)<} + :Br^- \longrightarrow \mathrm{-C(Br){-}C(H)-}$$

Hydrogen (reduction): alkenes, which are unsaturated compounds, add hydrogen to become saturated compounds:

$$H_2C{=}CH_2 + H_2 \longrightarrow H{-}CH_2{-}CH_2{-}H$$

Conditions Add hydrogen gas at high temperature and pressure, in the presence of a nickel catalyst

This reaction is used in the food industry during the manufacture of margarine.

Halogeno-compounds

Halogenoalkanes are compounds in which a **halogen atom** is bonded to a **saturated carbon atom**.

Reactions

Halogenoalkanes react with **aqueous sodium (or potassium) hydroxide**: a substitution reaction occurs, producing an alcohol.

Example

$$CH_3CH_2CH_2CH_2Br + OH^- \longrightarrow CH_3CH_2CH_2CH_2OH + Br^-$$

1-bromobutane butan-1-ol

Conditions Heat under reflux with aqueous sodium hydroxide solution

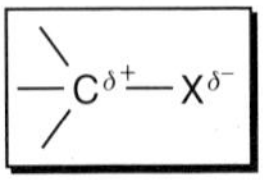

Because the carbon–halogen bond is polar the OH^- attacks the δ^+ carbon atom and so the reaction is classified as a **nucleophilic substitution**. Reactions in which an –OH group replaces a halogen atom are also called **hydrolysis reactions**.

$C^{\delta+}—X^{\delta-}$

The mechanism of nucleophilic substitution in primary halogenoalkanes proceeds as follows, using 1-bromobutane as an example:

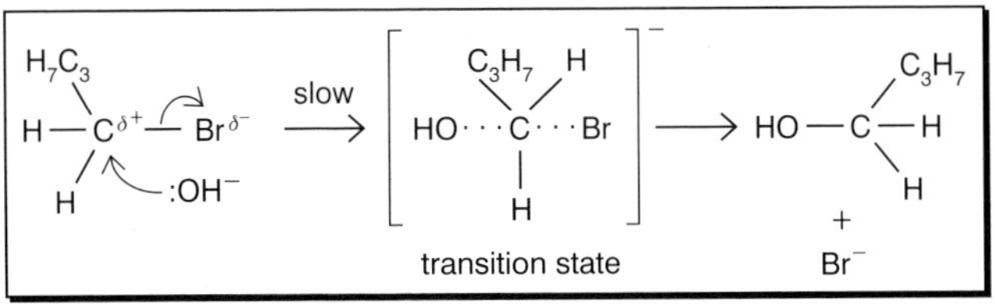

H_7C_3, H—$C^{\delta+}$—$Br^{\delta-}$, H, $:OH^-$ $\xrightarrow{\text{slow}}$ $[HO\cdots C(C_3H_7)(H)(H)\cdots Br]^-$ $\longrightarrow$ HO—C(C_3H_7)(H)—H + Br^-

transition state

Because there are two species in the slow step, the mechanism type is abbreviated to S_N2.

A transition state is a highly unstable species in which some bonds are partially broken and others are partially made.

A **nucleophile** is a species that attacks electron-deficient sites and **donates a lone pair** of electrons to form a covalent bond. Halogenoalkanes get more reactive as the C–X bond gets longer and therefore weaker. **Iodoalkanes** therefore react fastest and **fluoroalkanes** slowest.

With **ethanolic potassium hydroxide**, an **elimination reaction** occurs, producing an alkene.

Example

$CH_3CH_2CH_2CH_2Br + KOH \longrightarrow CH_3CH_2CH = CH_2 + KBr + H_2O$

1-bromobutane but-1-ene

Conditions Heat under reflux with a concentrated solution of potassium hydroxide in ethanol

An elimination reaction involves the removal of atoms from an organic molecule to form simple molecules, such as HBr, HCl and H_2O. The removal usually results in the formation of a multiple bond, such as $>C=C<$.

With **potassium cyanide**, a **nucleophilic substitution reaction** occurs. This increases the length of the carbon chain by one unit.

Example

$CH_3CH_2CH_2CH_2Br + KCN \longrightarrow CH_3CH_2CH_2CH_2CN + KBr$

1-bromobutane pentanenitrile

Conditions Heat under reflux with a solution of potassium cyanide in ethanol

The nucleophile is the cyanide ion, CN^-

With **ammonia**, a nucleophilic substitution reaction occurs.

Example

$CH_3CH_2CH_2CH_2Br + 2NH_3 \longrightarrow CH_3CH_2CH_2CH_2NH_2 + NH_4Br$

1-bromobutane butylamine

Conditions Heat with an excess of a concentrated solution of ammonia in ethanol under pressure in a sealed tube

Testing for halides

Once the halogen atom has left the halogenoalkane as the **halide ion**, X^-, silver nitrate solution can be used to identify the halide as in inorganic chemistry. The three essential steps are:

- warm the halogenoalkane with aqueous sodium hydroxide solution, e.g. $R{-}X + OH^- \longrightarrow R{-}OH + X^-$;
- add sufficient dilute nitric acid to neutralise any excess hydroxide ions;
- add aqueous silver nitrate solution.

Observations A **white** precipitate (soluble in dilute ammonia) indicates a **chloroalkane**. A **cream** precipitate (insoluble in dilute ammonia but soluble in concentrated ammonia) indicates a **bromoalkane**. A **yellow** precipitate (insoluble in both dilute and concentrated ammonia) indicates an **iodoalkane**.

Alcohols

Alcohols are compounds in which a **hydroxyl (–OH) group** is bonded to a **saturated carbon atom**.

Naming

The name of an alcohol is derived from the parent alkane; the letter 'e' is removed and replaced by '**ol**'. Where necessary, a number is added before the 'ol' to show the position of the hydroxyl group.

Examples

CH_3OH	methanol
$CH_3CH_2CH_2OH$	propan-1-ol
$(CH_3)_3COH$	2-methylpropan-2-ol

Alcohols containing two –OH groups are called **diols** and three –OH groups **triols**.

Examples

```
     H   H
     |   |
  H—C—C—H     ethane-1,2-diol
     |   |
    OH  OH
```

Alcohols in which the –OH group is attached to a carbon atom bonded to one or no alkyl groups is called a **primary** alcohol.

```
     H
     |
   —C—OH
     |
     H
```

If the –OH group is attached to a carbon atom with two other carbon atoms directly bonded to it, the alcohol is a **secondary** alcohol (**R = alkyl group**, e.g. –CH_3).

```
     H
     |
  R—C—OH
     |
     R′
```

If three alkyl groups are attached to the carbon atom bonded to the –OH group, the alcohol is a **tertiary** alcohol.

```
     R
     |
 R′—C—OH
     |
     R′′
```

Reactions

Primary and secondary alcohols undergo **oxidation** when reacted with a mixture of **acidified potassium dichromate (VI)** and **dilute sulphuric acid**. Primary alcohols are oxidised in the sequence

primary alcohol ⟶ aldehyde ⟶ carboxylic acid.

To obtain the **aldehyde**, further oxidation is prevented by distilling off the aldehyde as it is produced.

Example

```
     H   H                       H     O
     |   |                       |    //
  H—C—C—OH   + [O]  ⟶   H—C—C         + H2O
     |   |                       |    \
     H   H                       H     H
    ethanol                     ethanal
```

Reagents Potassium dichromate (VI) and dilute sulphuric acid

Conditions Distil off aldehyde as formed

The final oxidation product of a primary alcohol is a **carboxylic acid**.

Example

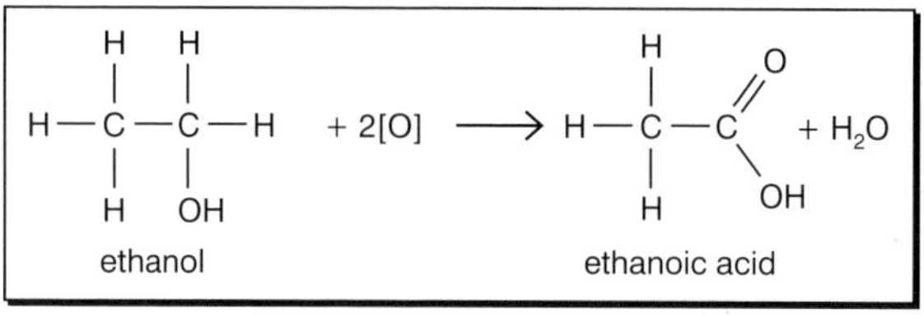

```
    H   H              H   O
    |   |              |  //
H — C — C — H + 2[O] ⟶ H — C — C     + H2O
    |   |              |   \
    H   OH             H    OH
  ethanol            ethanoic acid
```

Aldehydes, therefore, can also be oxidised to carboxylic acids.

Example

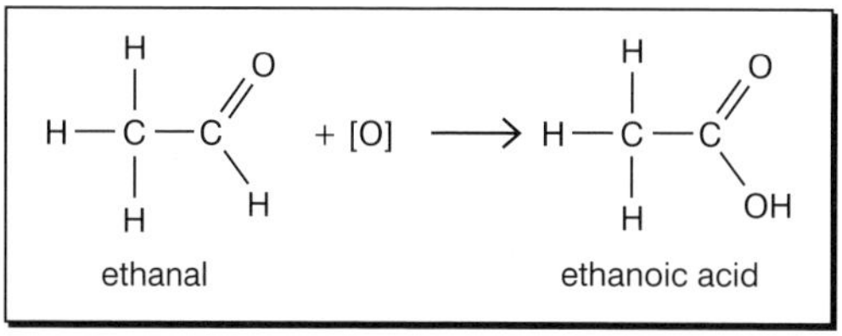

```
    H   O                 H   O
    |  //                 |  //
H — C — C     + [O] ⟶ H — C — C
    |   \                 |   \
    H    H                H    OH
  ethanal              ethanoic acid
```

For both these reactions:

Reagents $K_2Cr_2O_7$/dilute H_2SO_4

Conditions Heat under reflux

During these reactions, orange solutions of potassium dichromate (VI) are reduced to green solutions of chromium (III) ions. Secondary alcohols are oxidised to **ketones** when heated under reflux with the same reagents.

Example

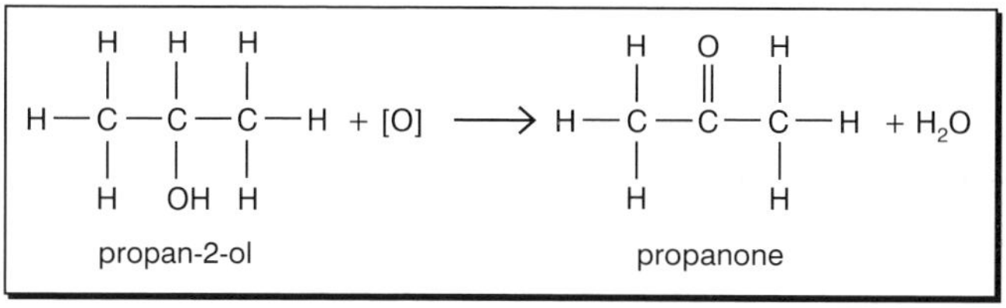

```
    H   H   H                  H   O   H
    |   |   |                  |   ‖   |
H — C — C — C — H + [O] ⟶ H — C — C — C — H + H2O
    |   |   |                  |       |
    H   OH  H                  H       H
   propan-2-ol                 propanone
```

Tertiary alcohols are resistant to oxidation under these conditions.

Alcohols form alkenes in an **elimination reaction** when reacted with **dehydrating** agents.

Example

$CH_3CH_2OH \longrightarrow H_2C{=}CH_2 + H_2O$
ethanol ethene

Suitable reagents and conditions Excess concentrated sulphuric acid (or concentrated phosphoric (V) acid) at 170°C *or* passing vapour of the alcohol over a heated catalyst of aluminium oxide

Alcohols undergo **nucleophilic substitution** at the δ^+ carbon atom when reacted with **halogenating agents**.

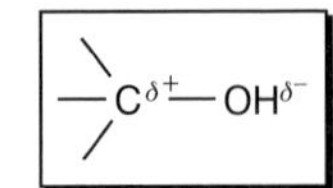

Example **Chlorination**

$CH_3CH_2OH + PCl_5 \longrightarrow CH_3CH_2Cl + POCl_3 + HCl$

ethanol chloroethane

Reagent: Phosphorus pentachloride, PCl_5

Conditions: At r.t.p.

This reaction is used to test for the –OH group because steamy fumes of the acidic gas hydrogen chloride are observed.

Example **Bromination**

$CH_3CH_2CH_2OH + HBr \longrightarrow CH_3CH_2CH_2Br + H_2O$

propan-1-ol 1-bromopropane

Reagents NaBr/concentrated H_2SO_4

Conditions Heat

Hydrogen bromide is prepared 'in situ' by heating sodium bromide and concentrated sulphuric acid:

$NaBr + H_2SO_4 \longrightarrow NaHSO_4 + HBr$

Example **Iodination**

$3C_2H_5OH + PI_3 \longrightarrow 3C_2H_5I + H_3PO_3$

ethanol iodoethane

Reagents Addition of an alcohol to red phosphorus and iodine

Conditions Heat under reflux

The phosphorus triiodide is formed 'in situ' by the reaction: $2P + 3I_2 \longrightarrow 2PI_3$.

Check yourself

1 Define the term 'homologous series'. (2)

2 Explain what is meant by:
  (a) structural isomers (2)
  (b) geometrical isomers (2)

3 Draw out and name all isomers of the compound C_5H_{12}. (3)

4 Define the following:
  (a) electrophile (1)
  (b) nucleophile (1)
  (c) free radical (1)
  (d) homolytic fission (1)
  (e) heterolytic fission (1)

5 Give the equation for the reaction between butane and one mole of chlorine gas. What are the conditions for the reaction? (2)

6 Give the equation for the reaction between propene and bromine. What are the conditions for the reaction? (2)

7 Give the equation for the reaction between 2-bromo-3-methylbutane and dilute aqueous potassium hydroxide. What are the conditions for the reaction? (3)

8 When 2-bromo-3-methylbutane is reacted with hot ethanolic potassium hydroxide, a mixture of two isomeric alkenes is formed. Give the structural formulae of the two alkenes. (2)

9 Classify the type of reaction occurring in:
  (a) Question 5 (2)
  (b) Question 6 (2)
  (c) Question 7 (2)
  (d) Question 8 (1)

10 Draw out the four structural isomers of the compound $C_4H_{10}O$ that are alcohols. (4)

11 Identify the alcohols, in your answers to Question 10, that would undergo oxidation when heated under reflux with a solution of potassium dichromate (VI) in dilute sulphuric acid. Give the structural formula of the final organic product of the reaction in each case. (6)

The answers are on page 119.

Applied organic chemistry

Chemicals from crude oil (petroleum)

Crude oil is formed by the long-term effects of heat and pressure on marine deposits. It is a complex mixture consisting mainly of alkane hydrocarbons.

The first step in the treatment of crude oil at a refinery is **fractional distillation**. The mixture is separated into various groupings of compounds, called fractions, by making use of the different boiling points of the components of the mixture. The main fractions obtained from the initial distillation of crude oil are shown below.

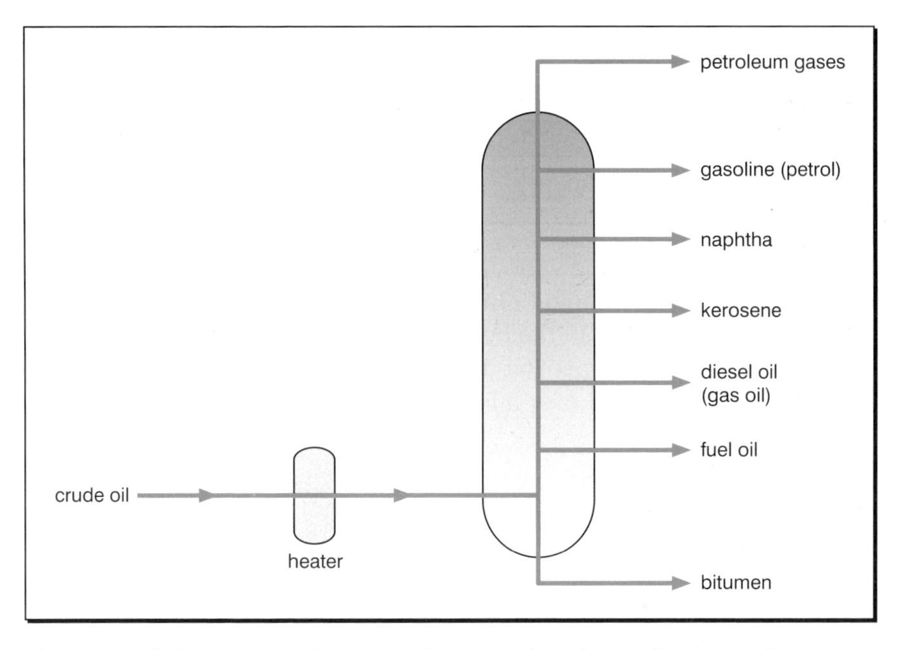

The uses of the various fractions from crude oil are shown in the table overleaf.

Name of fraction	Physical state	Boiling point range	Uses	Length of carbon chain
Petroleum gas	Gas	Up to 25°C	Calor gas; camping gas	C_1 to C_4
Gasoline (petrol)	Liquid	25–160°C	Petrol for cars	C_4 to C_{10}
Naphtha	Liquid	100–150°C	Manufacture of other petrochemicals	C_7 to C_{14}
Kerosene	Liquid	160–250°C	Jet fuel; heating fuel	C_{10} to C_{16}
Diesel oil (gas oil)	Liquid	250–300°C	Central heating fuel; fuel for cars, lorries and trains	C_{16} to C_{20}
Fuel oil	Liquid	Over 350°C	Fuel for ships and power stations; central heating fuel	C_{30} to C_{40}
Bitumen	Solid		Roofing; road surfaces	C_{50} and up

As the proportions of the various fractions produced do not match consumer demand, **catalytic cracking** is used to break down larger hydrocarbon molecules in the heavy fractions into smaller molecules, such as those used as gasoline. In industry, the vapour of the alkane being cracked is passed over a heated catalyst of aluminium oxide in the absence of air.

Example When the alkane called **decane** ($C_{10}H_{22}$) is cracked, the following reaction occurs: $C_{10}H_{22} \longrightarrow C_8H_{18} + C_2H_4$

decane octane ethene

The larger hydrocarbon molecule has been broken down into a mixture of a shorter chain alkane plus an alkene. Both of these products are useful: the alkane for the manufacture of petrol and the alkene for the manufacture of ethanol, by hydration, and plastics such as (poly)ethene.

Industrial production of ethanol

Much **ethanol** is manufactured by the **hydration** of ethene. The reaction is an **addition reaction** between steam and ethene at 300 °C, in the presence of a solid phosphoric acid catalyst, at a pressure of about 70 atmospheres.

$C_2H_{4(g)} + H_2O_{(g)} \longrightarrow C_2H_5OH_{(g)}$

Ethanol is also produced by the **fermentation** of sugars such as glucose. The reaction is carried out at about 35 °C in the presence of yeast, which contains an enzyme (biological catalyst) called **zymase**.

$C_6H_{12}O_{6(aq)} \longrightarrow 2C_2H_5OH_{(aq)} + 2CO_{2(g)}$

Air is excluded from the fermentation mixture (**anaerobic conditions**) to prevent the oxidation of ethanol to ethanoic acid. The process of fractional distillation is then used to obtain a more concentrated solution of ethanol.

Countries that have plenty of oil reserves, and are relatively rich, use the hydration method to produce ethanol, whereas those with a warm climate (where sugar can easily grow) and that are relatively poor, with no oil reserves, are more likely to use the fermentation method. In addition to its use in alcoholic beverages (fermentation method), ethanol is used industrially both as a fuel and as a solvent.

Addition polymers

Polymers can be formed from compounds containing a $>C=C<$ double bond. Alkenes, such as ethene, can undergo **addition polymerisation** to form a polymer. A polymer is a compound consisting of very long chain molecules built up from smaller molecular units, called **monomers**. The polymerisation of ethene, to form poly(ethene), is a **free radical addition reaction**.

$nC_2H_{4(g)} \longrightarrow (C_2H_4)_{n(s)}$

ethene — poly(ethene)

monomer — polymer

n is approximately 10 000.

The polymer molecules, therefore, have the same empirical formula as those of the monomer. The conditions for the reaction depend upon whether low-density or high-density poly(ethene) is required.

For **low-density poly(ethene)**, a high pressure (over 1500 atmospheres), a temperature of 200 °C and a trace of oxygen as a catalyst are needed.

For **high-density poly(ethene)**, a pressure of 2 to 6 atmospheres, a temperature of 60 °C and a catalyst of titanium (IV) chloride and triethyl aluminium are needed.

The structural formulae for the monomer and polymer are represented as follows.

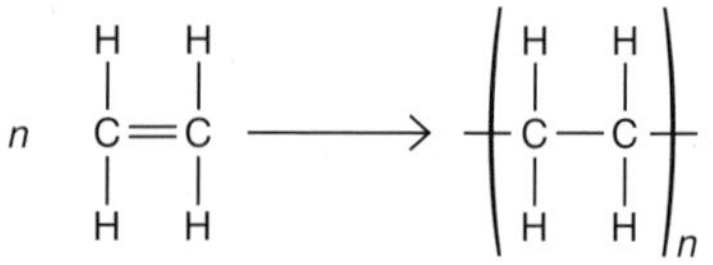

$$n\ \mathrm{H_2C{=}CH_2} \longrightarrow \left(\!-\mathrm{CH_2{-}CH_2}-\!\right)_n$$

ethene → poly(ethene)

This structure is called the **repeating unit** of the polymer poly(ethene):

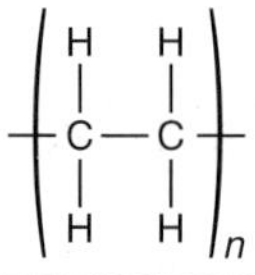

$$\left(\!-\mathrm{CH_2{-}CH_2}-\!\right)_n$$

Other polymers, based on compounds similar to ethene, can be formed. One or more of the hydrogen atoms in ethene is substituted by another atom or group.

Examples

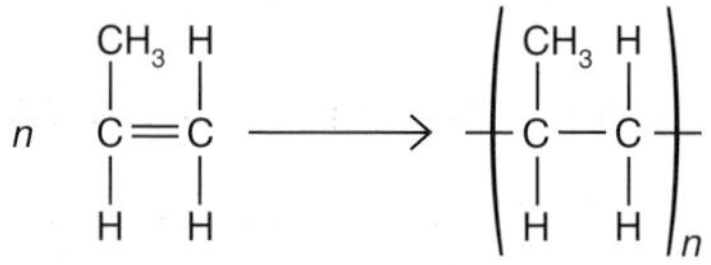

$$n\ \mathrm{CH_3CH{=}CH_2} \longrightarrow \left(\!-\mathrm{CH(CH_3){-}CH_2}-\!\right)_n$$

propene → poly(propene)

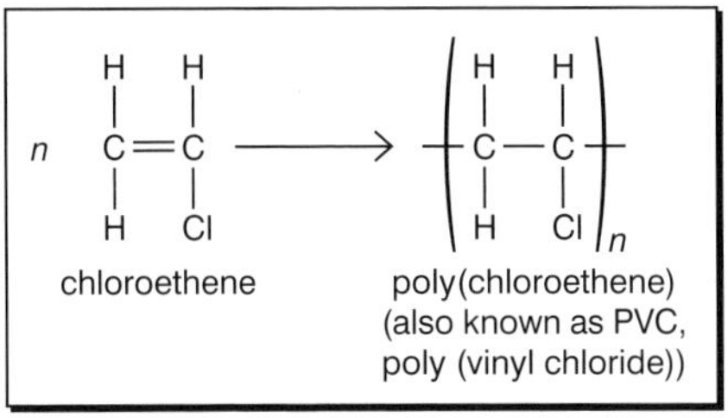

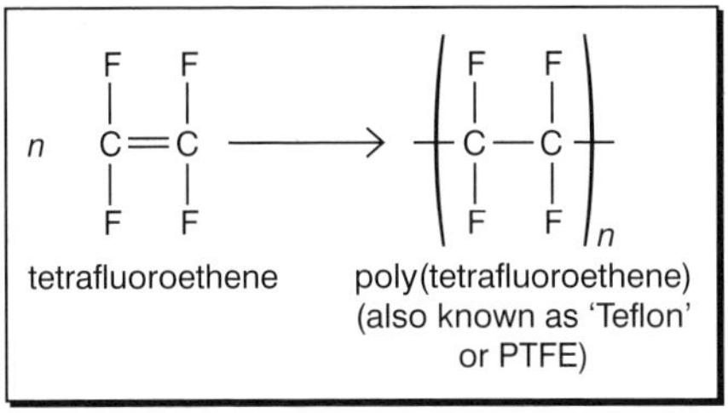

chloroethene → poly(chloroethene)
(also known as PVC, poly (vinyl chloride))

tetrafluoroethene → poly(tetrafluoroethene)
(also known as 'Teflon' or PTFE)

A table showing the uses of these polymers is below.

Name of polymer	Uses
Poly(ethene)	Plastic bags, food boxes, squeezy bottles, buckets, washing-up bowls
Poly(propene)	Ropes, carpets, clothing, pipes, crates
Poly(chloroethene)	Rainwear, plastic handbags, floor-tiles, guttering, electric cable, sails, window frames, drain pipes
Poly(tetrafluoroethene)	Non-stick ovenware, chemicals for seals and burette taps

Applied inorganic chemistry

Manufacture of ammonia by the Haber process

The effect of pressure, temperature and a catalyst on both the rate of a reaction and the position of an equilibrium is illustrated in the conditions employed for the manufacture of ammonia, NH_3.

Ammonia is synthesised from its elements nitrogen and hydrogen. The nitrogen is obtained by the fractional distillation of liquid air. The hydrogen is obtained by the reaction of methane (from natural gas) with steam.

$$CH_{4(g)} + H_2O_{(g)} \longrightarrow CO_{(g)} + 3H_{2(g)}$$

Carbon monoxide would 'poison' the catalyst used in the **Haber process** and is therefore removed from the system.

The nitrogen and hydrogen gases, in a 1:3 ratio, are passed over a heated iron catalyst at a temperature of 400°C and a pressure of about 200 atmospheres.

$$N_{2(g)} + 3H_{2(g)} \rightleftharpoons 2NH_{3(g)} \quad \Delta H = -92\,\text{kJ mol}^{-1}$$

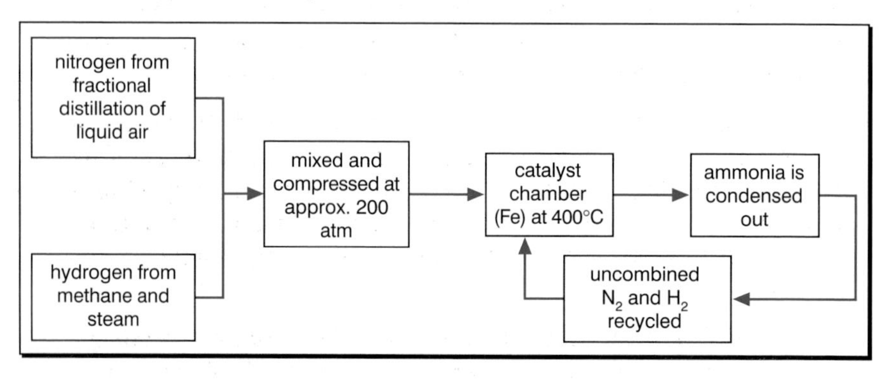

Because the reaction is an equilibrium process, with a typical yield of 15% ammonia per cycle, the gases leaving the converter are cooled whilst still under pressure. The ammonia is liquefied and removed, whilst the unreacted nitrogen and hydrogen are recycled to the converter.

The conditions used industrially for the Haber process are those that sustain the economic viability of its manufacture. Out of necessity, a high yield in a long time must be balanced against a low yield in a shorter time, whilst minimising energy costs. The conditions employed indicate a compromise between these opposing outcomes, as the graphs illustrate.

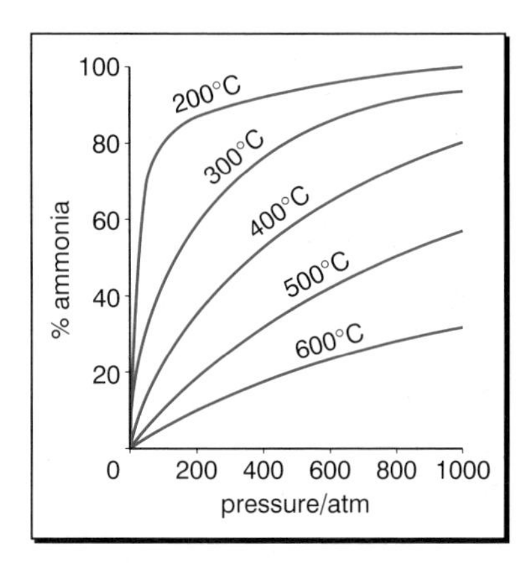

A **temperature** above 400 °C would increase the rate of the reaction, but would reduce the yield of ammonia. Below 400 °C, despite a higher yield, the rate of formation of ammonia is so slow as to be economically unviable.

A **pressure** of 200 atmospheres increases the rate of attainment of equilibrium and also increases the yield of ammonia. The use of much higher pressures is prohibited by the high costs of building suitable containers that are able to withstand greater pressure; in addition, the cost of the energy required to generate such large pressures is very expensive.

A **catalyst** of iron increases the rate of reaction, but has no effect on the percentage conversion to ammonia (the yield). The use of a catalyst allows the process to take place at a lower temperature than would otherwise be employed, so reducing the costs of the heat energy.

Manufacture of sulphuric acid

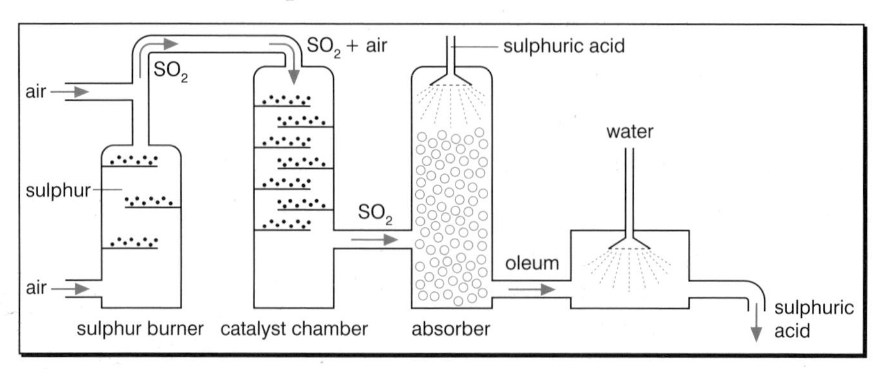

Sulphuric acid is an important industrial chemical. It has been said that the output of sulphuric acid is a measure of the wealth of a country, as it is used in the manufacture of fertilisers, detergents, pigments and fibres, amongst many other products. In the first stage of the process, sulphur is burned to produce sulphur dioxide.

$S_{(l)} + O_{2(g)} \longrightarrow SO_{2(g)}$

In the second stage, a mixture of sulphur dioxide and air is passed over a catalyst of vanadium (V) oxide, V_2O_5, at a temperature of about 430 °C. This is called the Contact process.

$2SO_{2(g)} + O_{2(g)} \rightleftharpoons 2SO_{3(g)}\ \Delta H = -196\,\text{kJ mol}^{-1}$

A high yield of sulphur trioxide is favoured by a high pressure (because there are fewer moles of gas molecules on the right-hand side) and a low temperature (the reaction is exothermic from left to right). In practice, a compromise temperature of 430 °C is used. At temperatures below this, the rate of reaction will be too slow and the catalyst is inactive. At higher temperatures, the yield of sulphur trioxide becomes uneconomically low. A pressure of 2 atmospheres, sufficient to ensure that the gases flow through the plant, is used because the costs required to operate at higher pressures cannot be justified.

In the final stage, the sulphur trioxide is absorbed in 98% sulphuric acid to form **oleum** (fuming sulphuric acid).

$$H_2SO_{4(l)} + SO_{3(g)} \longrightarrow \underset{\text{oleum}}{H_2S_2O_{7(l)}}$$

Oleum is then diluted with water to give sulphuric acid.

$$H_2S_2O_{7(l)} + H_2O_{(l)} \longrightarrow 2H_2SO_{4(l)}$$

Note that the sulphur trioxide is not absorbed directly by water, as the reaction is very exothermic and a corrosive mist of droplets of concentrated sulphuric acid is formed above the mixture. A double absorption sulphuric acid plant allows a 99.5% conversion of sulphur dioxide to sulphur trioxide to take place.

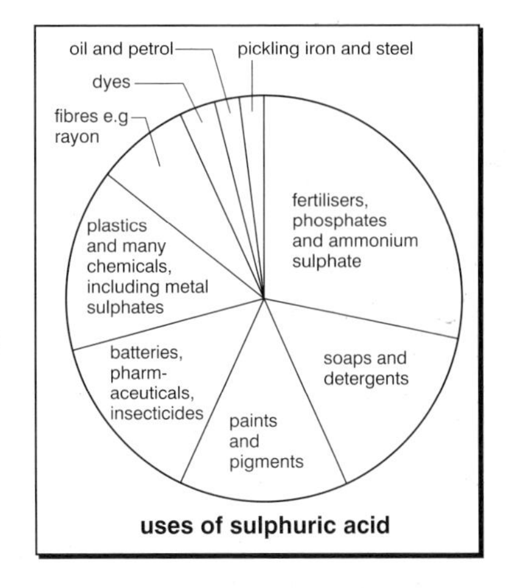

uses of sulphuric acid

Manufacture of inorganic fertilisers

Because nitrogenous fertilisers promote plant growth, ammonia solution (which is alkaline) is neutralised by dilute nitric acid to form ammonium nitrate.

$$NH_{3(aq)} + HNO_{3(aq)} \longrightarrow NH_4NO_{3(aq)}$$

Ammonium nitrate, e.g. ICI's 'NITRAM', is very widely used because of its high percentage by mass of nitrogen.

Sulphuric acid is used to manufacture other nitrogen-containing fertilisers such as ammonium sulphate, $(NH_4)_2SO_4$. Phosphorus-containing fertilisers, derived from rock phosphates such as $Ca_3(PO_4)_2$, are also manufactured using sulphuric acid.

Check yourself

1 Fractional distillation is used to separate crude oil into various fractions. What property of the fractions allows them to be separated in the column? (1)

2 A gas oil fraction from the distillation of crude oil contains hydrocarbons in the C_{16} to C_{20} range. These hydrocarbons can be cracked by heating in the presence of a catalyst.

(a) Give the molecular formula of the alkane containing 19 carbon atoms. (1)

(b) On cracking the alkane $C_{16}H_{34}$, the products include ethene and propene in the mole ratio of 2:1, plus one other compound. Give the equation for the cracking reaction described. (2)

(c) Use your answer to (b) to explain the term 'cracking'. (2)

3 Give an equation in each case for the industrial production of ethanol.

(a) By the fermentation of glucose. (1)

(b) By the hydration of ethene. (1)

4 The plastic poly(chloroethene), PVC, is made from the monomer chloroethene, C_2H_3Cl, by a polymerisation reaction.

(a) What type of reaction is the formation of poly(chloroethene) from chloroethene? (2)

(b) Draw the full structural formula of chloroethene. (1)

(c) Draw the structure of part of the poly(chloroethene) molecule containing six carbon atoms. (1)

(d) Show the repeating unit of the poly(chloroethene) polymer. (1)

5 Polymers such as PVC and PTFE ('Teflon') are very stable. Use the bond enthalpy values $\overline{E}(C\text{–}Cl) = +346\ kJ\ mol^{-1}$ and $\overline{E}(C\text{–}F) = +480\ kJ\ mol^{-1}$ to explain this fact. (2)

The answers are on page 121.

Check yourself answers

ATOMIC STRUCTURE (page 10)

1

Particle	Relative mass	Relative charge
Proton	1	1+ (1)
Neutron	1	0 (1)
Electron	$\frac{1}{1840} \approx 0$	1− (1)

Relative mass has no units. Quote a value between $\frac{1}{1800}$ to $\frac{1}{2000}$ for the relative mass of an electron.

2 Number of protons plus number of neutrons (or the number of nucleons) in a nucleus (1). Avoid 'number of protons and neutrons'.

3 Number of protons in the nucleus / in the atom (1). Avoid 'number of electrons' and avoid 'number of protons in the element'.

4 Average mass (1) of an atom compared with $\frac{1}{12}$th of the mass of an atom of ^{12}C (1). Mention of carbon-12 is essential.

5 Atoms of the same element (accept: same atomic number / proton number) (1) that differ only in the number of neutrons (accept: different mass number / isotopic mass) (1).

6 $A_r = \left(24 \times \frac{78.6}{100}\right) + \left(25 \times \frac{10.1}{100}\right) + \left(26 \times \frac{11.3}{100}\right)$ (1)
$A_r = 24.327$ (1)
$A_r = 24.3$ (to 3 significant figures) (1). A_r has NO units. Round up the answer to the required degree of accuracy, in this case, three significant figures.

7 **(a)** By an electric field (accept: negatively charged plate)(1).
(b) By a magnetic field (accept: magnets) (1). Learn the function of each part of the mass spectrometer.

8 As the atom radii increase down group 1, the outermost electron is further from the nucleus (1) **and** more shielded from the nuclear charge, so is held less tightly (despite the increasing number of protons in the nucleus of each atom) (1). Always answer in terms of factors such as distance, shielding and nuclear charge.

9 **(a)** $1s^2\ 2s^2\ 2p^5$ (1)
(b) $1s^2\ 2s^2\ 2p^6\ 3s^2\ 3p^6$ (1). In (b) the Cl^- ion is asked for. This species has 17 protons and 18 electrons.

10 Much more heat energy is released (1) when electrostatic forces of attraction between oppositely charged ions (Mg^{2+} and O^{2-}) are made (1). 'Suggest' questions require 'lateral thinking' rather than 'recall'.

Check yourself answers

FORMULAE, MOLES AND EQUATIONS (page 24)

1 **(a)** 12.0 g of C contains 6.0×10^{23} C atoms (1)

$\downarrow \div 12$ $\qquad\qquad \downarrow \div 12$

1.00 g of C contains $\frac{6.0 \times 10^{23}}{12.0}$ C atoms

$\downarrow \times 0.120$ $\qquad\qquad \downarrow \times 0.120$

0.120 g of C contains $\frac{0.120 \times 6.0 \times 10^{23}}{12.0}$ C atoms

$= 6.0 \times 10^{21}$ C atoms (1)

(b) Molar mass of $SO_2 = 32 + 32 = 64\ g\ mol^{-1}$

64.0 g of SO_2 contains 6.0×10^{23} SO_2 molecules (1)

$\downarrow \div 2$ $\qquad\qquad \downarrow \div 2$

32.0 g of SO_2 contains 3.0×10^{23} SO_2 molecules (1)

(c) Molar mass of $Na_2SO_4 = 46 + 32 + 64 = 142\ g\ mol^{-1}$

$Na_2SO_{4(aq)} \longrightarrow 2Na^+_{(aq)} + SO_4{}^{2-}{}_{(aq)}$

1 mole of $Na_2SO_4 \longrightarrow$ 2 moles of Na^+ ions

142 g $Na_2SO_4 \longrightarrow 2 \times 6.0 \times 10^{23}$

$= 12 \times 10^{23}$ Na^+ ions (1)

$\downarrow \div 10$ $\qquad\qquad \downarrow \div 10$

14.2 g $Na_2SO_4 \longrightarrow 1.2 \times 10^{23}$ Na^+ ions (1)

Make it clear whether you are referring to moles of atoms, molecules or ions. Note the importance of correct recall of formulae in part (c).

2 **(a)** Number of moles of atoms of O $= \frac{\text{mass of oxygen atoms}}{\text{molar mass of oxygen}}$

$= \frac{16.0\ g}{16.0\ g\ mol^{-1}}$ (1)

$= 1.00$ mol of O atoms (1)

(b) Number of moles of atoms N $= \frac{\text{mass of nitrogen atoms}}{\text{molar mass of nitrogen}}$

$= \frac{0.140\ g}{14.0\ g\ mol^{-1}}$ (1)

$= 0.0100$ mol of N atoms (1)

(c) Number of moles of atoms of Ag $= \frac{\text{mass of silver atoms}}{\text{molar mass of silver}}$

$= \frac{5.40\ g}{108\ g\ mol^{-1}}$ (1)

$= 0.0500$ mol of Ag atoms (1)

Notice how units can confirm that the formula for the number of moles of atoms has been used correctly.

3 **(a)** Mass of oxygen (g) $= 0.500 \text{ (mol)} \times 16.0 \text{ (g mol}^{-1}\text{)}$
$= 8.00$ g of oxygen atoms (1)

(b) Mass of sodium (g) $= 10.0 \text{ (mol)} \times 23.0 \text{ (g mol}^{-1}\text{)}$
$= 230$ g of sodium atoms (1)

(c) Mass of hydrogen (g) $= 0.0100 \text{ (mol)} \times 1 \text{ (g mol}^{-1}\text{)}$
$= 0.0100$ g of hydrogen atoms (1)

For each part, the formula: number of moles of atoms of an element $= \frac{\text{mass of substance}}{\text{molar mass}}$ has been rearranged to mass of substance = number of moles of atoms of an element × molar mass

4 **(a)** Molar mass of Br_2 $= 2 \times 80$
$= 160 \text{ g mol}^{-1}$ (1)

(b) Molar mass of HNO_3 $= 1 + 14 + (3 \times 16)$
$= 63 \text{ g mol}^{-1}$ (1)

(c) Molar mass of $CuSO_4 \cdot 5H_2O = 64 + 32 + (4 \times 16) + (5 \times 18)$
$= 250 \text{ g mol}^{-1}$ (1)

Remember to add up all the individual atomic masses, scaling up where appropriate.

5 **(a)** Number of moles of oxygen, O_2(mol) $= \frac{128 \text{ (g)}}{32.0 \text{ (g mol}^{-1}\text{)}}$
$= 4.00$ mol of O_2 (1)

(b) Number of moles of potassium nitrate, KNO_3 (mol) $= \frac{25.25 \text{ (g)}}{101 \text{ (g mol}^{-1}\text{)}}$
$= 0.250$ mol of KNO_3 (1)

(c) Number of moles of ethanol, C_2H_5OH $= \frac{414 \text{ (g)}}{46 \text{ (g mol}^{-1}\text{)}}$
$= 9.0$ mol of C_2H_5OH (1)

This question requires use of the general expression: number of moles of a substance (mol) $= \frac{\text{mass of sample (g)}}{\text{molar mass (g mol}^{-1}\text{)}}$

6 **(a)** Mass of sulphur dioxide (g) $= 2.00 \text{ (mol)} \times 64 \text{ (g mol}^{-1}\text{)}$
$= 128$ g of sulphur dioxide (1)

(b) Mass of sulphuric acid (g) $= 20.0 \text{ (mol)} \times 98 \text{ (g mol}^{-1}\text{)}$
$= 1960$ g of sulphuric acid (1)

(c) Mass of sodium hydroxide (g) $= 0.500 \text{ (mol)} \times 40 \text{ (g mol}^{-1}\text{)}$
$= 20$ g of sodium hydroxide (1)

For this question, the formula: number of moles of a substance $= \frac{\text{mass of substance}}{\text{molar mass}}$ has been rearranged to mass of substance = number of moles of a substance × molar mass.

7 (a)

	Fe : O
Mass ratio/g	1.12 : 0.48 (1)
Mole ratio/mol	$\frac{1.12}{56} : \frac{0.48}{16}$
=	0.02 : 0.03
Divide by smallest number	$\frac{0.02}{0.02} : \frac{0.03}{0.02}$
=	1 : 1.5
=	2 : 3 (1)

Therefore, empirical formula is Fe_2O_3

Remember to subtract the mass of iron from the mass of iron oxide, to calculate the mass of oxygen reacting. Note that a mole ratio of 1 : 1.5 is not rounded up to 1 : 2

(b)

	Na : C : O
Mass ratio/g in 100 g	43.4 : 11.3 : 45.3
Mole ratio/mol	$\frac{43.4}{23.0} : \frac{11.3}{12.0} : \frac{45.3}{16.0}$
=	1.887 : 0.942 : 2.831 (1)
Divide by smallest number	$\frac{1.887}{0.942} : \frac{0.942}{0.942} : \frac{2.831}{0.942}$
=	2 : 1 : 3 (1)

Therefore, empirical formula is Na_2CO_3

(c)

	C : H
Mass ratio/g in 100 g	82.75 : 17.25
Mole ratio/mol	$\frac{82.75}{12} : \frac{17.25}{1}$
=	6.896 : 17.25 (1)
Divide by smallest number	$\frac{6.896}{6.896} : \frac{17.25}{6.896}$
=	1 : 2.5
=	2 : 5 (1)

Therefore, empirical formula is C_2H_5

In parts (b) and (c), the percentage by mass tells you the mass ratio in a 100 g sample. In part (c), a ratio of 1 : 2.5 is rounded up to 2 : 5 (not 1 : 3).

8 Molar mass of the compound = $n \times$ empirical mass of the compound. The molar mass of the compound is given as 58 g mol^{-1}. The empirical mass of the compound of empirical formula C_2H_5 is $(12 \times 2 + 1 \times 5)$ = 29 g mol^{-1}. Therefore, to calculate n

$$n = \frac{\text{molar mass of the compound}}{\text{empirical mass of the compound}}$$

$= \frac{58 \text{ g mol}^{-1}}{29 \text{ g mol}^{-1}}$

$= 2$ (1)

It follows that the molecular formula of the compound is $(C_2H_5) \times 2 = C_4H_{10}$ (1). A quick check confirms that the molar mass is correct: molar mass of C_4H_{10} in g mol^{-1} = $(12 \times 4) + (1 \times 10) = 58$ g mol^{-1}

9 **(a)** Start with the full equation:

$AgNO_{3(aq)} + NaCl_{(aq)} \longrightarrow AgCl_{(s)} + NaNO_{3(aq)}$

Write the ions separately where appropriate:

$Ag^+_{(aq)} + NO_3^-{}_{(aq)} + Na^+_{(aq)} + Cl^-_{(aq)} \longrightarrow AgCl_{(s)} + Na^+_{(aq)} + NO_3^-{}_{(aq)}$ (1)

Delete the spectator ions to produce the ionic equation:

$Ag^+_{(aq)} + Cl^-_{(aq)} \longrightarrow AgCl_{(s)}$ (1)

(b) $Zn_{(s)} + H_2SO_{4(aq)} \longrightarrow ZnSO_{4(aq)} + H_{2(aq)}$

$Zn_{(s)} + 2H^+_{(aq)} + SO_4^{2-}{}_{(aq)} \longrightarrow Zn^{2+}_{(aq)} + SO_4^{2-}{}_{(aq)} + H_{2(g)}$ (1)

$Zn_{(s)} + 2H^+_{(aq)} \longrightarrow Zn^{2+}_{(aq)} + H_{2(g)}$ (1)

(c) $HCl_{(aq)} + NaOH_{(aq)} \longrightarrow NaCl_{(aq)} + H_2O_{(l)}$

$H^+_{(aq)} + Cl^-_{(aq)} + Na^+_{(aq)} + OH^-_{(aq)} \longrightarrow Na^+_{(aq)} + Cl^-_{(aq)} + H_2O_{(l)}$ (1)

$H^+_{(aq)} + OH^-_{(aq)} \longrightarrow H_2O_{(l)}$ (1)

These ionic equations summarise the following types of reaction: (a) the **precipitation** reaction between aqueous silver and aqueous chloride ions; (b) the **redox** reaction between zinc metal and dilute acid; (c) the **neutralisation** reaction between an acid and an alkali.

10 $Mg_{(s)} + Cl_{2(g)} \longrightarrow MgCl_{2(s)}$

1 mole $Mg_{(s)} \longrightarrow$ 1 mole $MgCl_{2(s)}$ (1)

24 g $Mg_{(s)} \longrightarrow$ 95 g $MgCl_{(s)}$

↓ × 2 ↓ × 2

48 g $Mg_{(s)} \longrightarrow$ 190 g $MgCl_{2(s)}$ (1)

11 $CaCO_{3(s)} \longrightarrow CaO_{(s)} + CO_{2(g)}$

1 mole $CaCO_{3(s)} \longrightarrow$ 1 mole $CaO_{(s)}$ (1)

100 g $CaCO_{3(s)} \longrightarrow$ 56 g $CaO_{(s)}$

↓ ÷ 56 ↓ ÷ 56

$\frac{100}{56}$ g $CaCO_{3(s)} \longrightarrow$ 1 g $CaO_{(s)}$

↓ × 14 ↓ × 14

$\frac{14 \times 100}{56}$ g $CaCO_{3(s)} \longrightarrow$ 14 g $CaO_{(s)}$

$\frac{14 \times 100}{56}$ tonnes $CaCO_{3(s)} \longrightarrow$ 14 tonnes $CaO_{(s)}$

= 25 tonnes $CaCO_{3(s)} \longrightarrow$ 14 tonnes $CaO_{(s)}$ (1)

Note that mass ratios are identical, whether measured in grams or tonnes.

12 Number of moles of H_2SO_4 = $\frac{9.80}{98}$ = 0.10

Number of moles of $NaNO_3$ = $\frac{8.50}{85}$ = 0.10

Number of moles of HNO_3 = $\frac{6.30}{63}$ = 0.10 (1)

Therefore, 1 mol H_2SO_4 reacts with 1 mol $NaNO_3$ to produce 1 mol HNO_3

$H_2SO_{4(l)} + NaNO_{3(s)} \longrightarrow HNO_{3(l)}$ (1)

The co-product, if the equation is to balance for mass, is $NaHSO_{4(s)}$. The full equation is, therefore:

$H_2SO_{4(l)} + NaNO_{3(s)} \longrightarrow HNO_{3(l)} + NaHSO_{4(s)}$ (1)

The ability to balance chemical equations is vital for AS Chemistry.

13 $Mg_{(s)} + H_2SO_{4(aq)} \longrightarrow MgSO_{4(aq)} + H_{2(g)}$

1 mole $Mg_{(s)}$ $\longrightarrow$ 1 mole of $H_{2(g)}$ (1)

24 g $Mg_{(s)}$ $\longrightarrow$ 24000 cm^3 $H_{2(g)}$

$\downarrow \div 2$ $\downarrow \div 2$

12 g $Mg_{(s)}$ $\longrightarrow$ 12000 cm^3 $H_{2(g)}$ (or 12 dm^3 $H_{2(g)}$) (1)

14 $2\ C_8H_{18(g)} + 25\ O_{2(g)} \longrightarrow 16\ CO_{2(g)} + 18\ H_2O_{(l)}$

2 moles $C_8H_{18(g)}$ + 25 moles $O_{2(g)}$ $\longrightarrow$ 16 moles $CO_{2(g)}$ + 18 moles $H_2O_{(l)}$

By Gay–Lussac's law:

2 volumes $C_8H_{18(g)}$ + 25 volumes $O_{2(g)}$ $\longrightarrow$ 16 volumes $CO_{2(g)}$

It follows that:

2 dm^3 $C_8H_{18(g)}$ + 25 dm^3 $O_{2(g)}$ $\longrightarrow$ 16 dm^3 $CO_{2(g)}$

Therefore:

(a) 25.0 dm^3 of oxygen are required to burn 2.00 dm^3 octane completely (1).

(b) 16.0 dm^3 of carbon dioxide are produced when 2.00 dm^3 octane are completely burned (1).

Water is produced as a liquid at the temperature at which the volumes are measured, therefore Gay–Lussac's law is not applicable to the water.

15 $KOH_{(aq)} + HNO_{3(aq)} \longrightarrow KNO_{3(aq)} + H_2O_{(l)}$

Moles of $HNO_{3(aq)}$ reacting = $\frac{0.100\ \text{mol dm}^{-3} \times 28.5\ \text{cm}^3}{1000}$

= 0.00285 mol $HNO_{3(aq)}$ (1)

From the above equation, 25.0 cm^3 of the $KOH_{(aq)}$ must also contain 0.00285 mol. Therefore:

concentration of $KOH_{(aq)}$ = $\frac{\text{moles of } KOH_{(aq)}\ (\text{mol}) \times 1000}{\text{volume of } KOH_{(aq)}\ (\text{cm}^3)}$

= $\frac{0.00285 \times 1000}{25.0}$

= 0.114 mol dm^{-3} (1)

Check yourself answers

FORMULAE, MOLES AND EQUATIONS (page 26)

16 $KOH_{(aq)} + HCl_{(aq)} \longrightarrow KCl_{(aq)} + H_2O_{(l)}$

Note that, in this equation, the concentration of the $KOH_{(aq)}$ is expressed in g dm^{-3}.

Moles of $KOH_{(aq)}$ reacting

$$= \frac{\text{concentration of KOH (mol dm}^{-3}) \times \text{volume of KOH (cm}^3)}{1000}$$

In 5.6 g KOH there is $\frac{5.6 \text{ g}}{56 \text{ g mol}^{-1}} = 0.10$ mol KOH

because the molar mass of KOH is 56 g mol^{-1} . Therefore the concentration of the KOH in mol dm^{-3} is 0.10 mol dm^{-3} (1).

Moles of $KOH_{(aq)}$ reacting $= \frac{0.10 \times 40.0}{1000}$

$= 0.0040$ mol KOH (1)

The concentration of the $HCl_{(aq)}$ $= \frac{\text{moles of HCl reacting} \times 1000}{\text{volume of HCl}_{(aq)}}$

$= \frac{0.0040 \times 1000}{40.0}$

$= 0.10$ mol dm^{-3} (1)

17 $C_6H_{6(l)} + HNO_{3(l)} \longrightarrow C_6H_5NO_{2(l)} + H_2O_{(l)}$

From the equation:

1 mole $C_6H_{6(l)} \longrightarrow$ 1 mole $C_6H_5NO_{2(l)}$

78 g $C_6H_{6(l)} \longrightarrow$ 123 g $C_6H_5NO_{2(l)}$

39.0 g $C_6H_{6(l)}$ should give 61.5 g $C_6H_5NO_{2(l)}$ (1)

Therefore, percentage yield $= \frac{\text{actual mass of product}}{\text{calculated mass of product}} \times 100$

$= \frac{49.2}{61.5} \times 100$

$= 80.0\%$ (1)

STRUCTURE AND BONDING (page 37)

1 To melt the giant structure of diamond, very strong covalent bonds (1) between carbon atoms (1) must be broken. In contrast, only weak van der Waals' forces (1) between iodine (I_2) molecules (1) need to be broken when iodine is heated beyond its melting temperature. The strong covalent bonds within the iodine molecules are not affected.

2 To melt sodium chloride, very strong (1) electrostatic forces of attraction between positive and negative ions (1) must be overcome. Do not just write 'strong bonds'.

3 The bonding in a metal such as aluminium consists of Al^{3+} cations (1) in a 'sea' of delocalised electrons (1). Therefore, current can flow due to the presence of mobile electrons (three per Al atom) (1).

4 The Mg^{2+} cation (1) (do not put 'magnesium' or 'Mg') is more able to polarise the larger iodide, I^-, anion (1) than the smaller Cl^- ion (1).

5 The power of an atom (1) to attract the bonding electrons in a covalent bond (1). Do not confuse this with electron affinity.

6 **(a)** Giant molecular = SiO_2 (1)
(b) Simple molecular = CO_2 (1)
(c) Giant ionic = K^+Br^- (1)
(d) Giant metallic = Cu (1)

7 **(a)** Despite the molar mass of HCl, and hence the number of e^- in the molecule, being greater than that of HF, HF has the higher boiling temperature due to the presence of hydrogen bonds (1). Hydrogen bonds are stronger intermolecular forces than van der Waals' forces (1). More heat energy is required, therefore, to break hydrogen bonds.

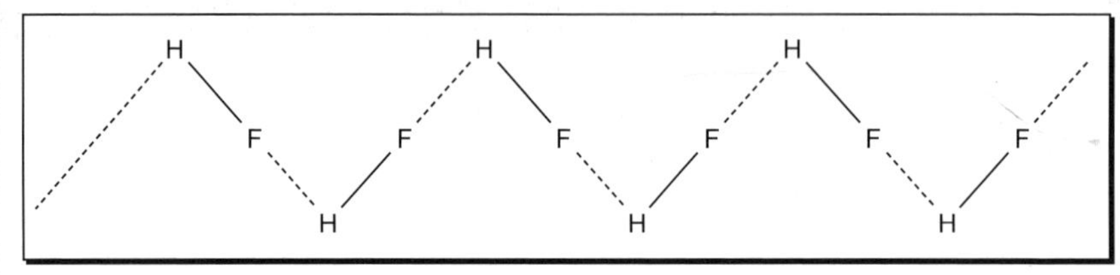

(b) The boiling temperatures increase along the series HCl $\longrightarrow$ HI, because the number of electrons in each molecule increases (1). The van der Waals' forces, therefore, increase in strength along the series HCl $\longrightarrow$ HI (1), so more energy is required to separate HI molecules than HCl molecules in the liquid state.

8 **(a)** Six bond pairs around S, no lone pairs (1). To maximise the separation between electron pairs, an octahedral shaped molecule is formed (1).

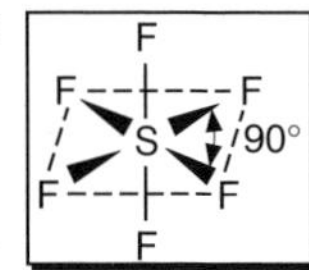

(b) The carbonate ion has one C=O double bond and two C–O⁻ single bonds. Because multiple bonds are treated as single bonds, the shape of the ion is trigonal planar (1), as if three bond pairs (1).

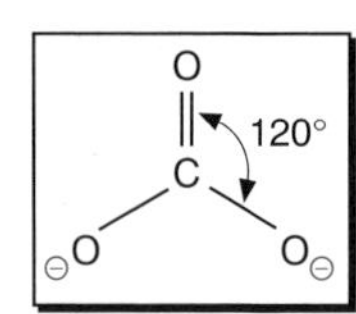

Check yourself answers

STRUCTURE AND BONDING (page 37)

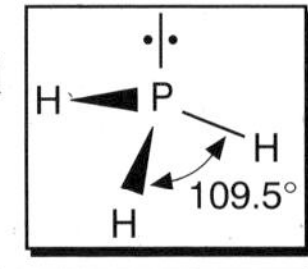

(c) Three bond pairs and one lone pair (1). To maximise the separation between the electron pairs, a trigonal pyramidal shape is adopted (1). H P H bond angle <109.5°, because lone pair/bond pair repulsion is greater than bond pair/bond pair repulsion.

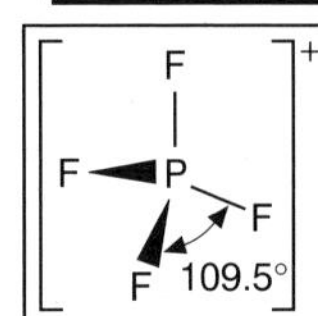

(d) As the species is positively charged, there are only four bond pairs around the central P atom (1). The shape of the ion is tetrahedral (1).

REDOX REACTIONS (page 44)

1 **(a)** 0 (1)
(b) 0 (1) In (a) and (b), atoms are in uncombined elements.
(c) +4 (1) (2N) + (4 × −2) = 0; N = +4
(d) +5 (1) (N) + (3 × −2) = −1; N = +5
(e) +7 (1)
(f) +3 (1)
(g) −1 (1) This is unusual; O in a peroxide, therefore oxidation number −1
(h) +4 (1)
(i) +2 (1) F stays at −1, as more electronegative than O
(j) +2.5 (1) Oxidation number can be a non-integer when it is an average value

2 **(a)** Chromium (III) fluoride (1)
(b) Iron (III) hydroxide (1)
(c) Iron (II) sulphide (1)
(d) Manganese (II) carbonate (1)
(e) Copper (II) sulphate – 5 – water (1) Note the water of crystallisation

3 **(a)** I: from −1 to 0; oxidation (1)
Cl: from 0 to −1; reduction (1)
(b) Fe: from +3 to +2; reduction (1)
Sn: from +2 to +4; oxidation (1)
(c) Ca: from 0 to +2; oxidation (1)
H: from 0 to −1; reduction (1)

Note the increase in oxidation number is oxidation; decrease in oxidation number is reduction.

Check yourself answers

Redox Reactions (page 44)

4 **(a)** oxidation: $2I^-_{(aq)} \longrightarrow I_{2(aq)} + 2e^-$ (1)
reduction: $Cl_{2(aq)} + 2e^- \longrightarrow 2Cl^-_{(aq)}$ (1)
(b) oxidation: $Sn^{2+}_{(aq)} \longrightarrow Sn^{4+}_{(aq)} + 2e^-$ (1)
reduction: $Fe^{3+}_{(aq)} + e^- \longrightarrow Fe^{2+}_{(aq)}$ (1)
(c) oxidation: $Ca_{(s)} \longrightarrow Ca^{2+}_{(s)} + 2e^-$ (1)
reduction: $H_{2(g)} + 2e^- \longrightarrow 2H^-_{(s)}$ (1)

Oxidation: loss of electrons; reduction: gain of electrons (OIL RIG).

5 **(a)** $H_2O_{2(aq)} + H_2S_{(g)} \longrightarrow S_{(s)} + 2H_2O_{(l)}$ (1). Add the half-equations, as number of electrons in each reaction is the same.
(b) $I_{2(aq)} + 2e^- + 2S_2O_3^{2-}{}_{(aq)} \longrightarrow S_4O_6^{2-}{}_{(aq)} + 2I^-_{(aq)} + 2e^-$
Overall: $I_{2(aq)} + 2S_2O_3^{2-}{}_{(aq)} \longrightarrow S_4O_6^{2-}{}_{(aq)} + 2I^-_{(aq)}$ (1). Multiply second equation by 2, then add together.
(c) $2MnO_4^-{}_{(aq)} + 16H^+_{(aq)} + 10e^- + 5H_2O_{2(aq)} \longrightarrow$
$2Mn^{2+}_{(aq)} + 8H_2O_{(l)} + 5O_{2(g)} + 10H^+_{(aq)} + 10e^-$
Overall: $2MnO_4^-{}_{(aq)} + 6H^+_{(aq)} + 5H_2O_{2(aq)} \longrightarrow 2Mn^{2+}_{(aq)} + 8H_2O_{(l)} + 5O_{2(g)}$ (1). Multiply first equation by 2 and second equation by 5, then add together. Note that e^- and some $H^+_{(aq)}$ can be cancelled from both sides.

The Periodic Table (page 54)

1 The melting temperatures of the metals Na to Al increase (1) as the strength of the metallic bonding increases (1). Silicon has the highest melting temperature (1) as there are very strong covalent bonds between atoms (1). The melting temperatures of the elements phosphorus to argon are much lower (1) as only weak van der Waals' forces (1) exist between molecules.

2 As we move across the period, atomic radius decreases as electrons are added to the same shell ($n = 3$) (1) and, therefore, the outermost electrons experience the same shielding (1) from the nuclear charge. The increase in the number of protons (1) causes the decrease in radius from Na to Ar. These factors also explain the general increase in first ionisation energy across the period.

3 **(a)** Lilac (1)
(b) Yellow (1)
(c) Carmine red (1) Not just 'red'.

4 Heat energy produces atoms of the metal with electrons in a higher energy level shell than the ground state (1). Energy in the form of visible light is emitted when electrons fall back down to their normal shells (1).

5 **(a)** $Ca_{(s)} + 2H_2O_{(l)} \longrightarrow Ca(OH)_{2(aq)} + H_{2(g)}$ Species and balancing (1). Correct state symbols (1).
(b) As the atomic mass of the metal increases, the group 2 hydroxides become more (1) soluble in water.

6 $2Na_{(s)} + O_{2(g)} \longrightarrow Na_2O_{2(s)}$ (1)

Check yourself answers

THE PERIODIC TABLE (page 54)

$2Mg_{(s)} + O_{2(g)} \longrightarrow 2MgO_{(s)}$ (1)
The larger, less polarising, Na^+ cation is able to accommodate enough of the bigger peroxide ions (O_2^{2-}) to form a stable lattice (1). The smaller, more polarising, Mg^{2+} ion forms a more stable lattice with the O^{2-} ion (1).

7 $2Mg(NO_3)_{2(s)} \longrightarrow 2MgO_{(s)} + 4NO_{2(g)} + O_{2(g)}$ Species (1) and balancing (1).

8 As group 2 is descended, the carbonates become more stable to heat (1). The trend is the same for the nitrates as well. The ionic radii of the group 2 cations (M^{2+}) increase down the group (1) and so the ability of the cation to distort the electron cloud around the anion gets less (1). Alternative: as the charge density of the cation decreases down the group, the polarising power of the cation gets less.

9 **(a)** Steamy fumes (1).
(b) Reddy-brown vapour (1).
(c) Purple vapour (1).
(d) As the halogen atom increases in size, the H–X bond becomes longer and weaker (1), therefore the ease of oxidation of the hydrogen halides is: HI > HBr > HCl (1). The order of their reducing power is, therefore, the same!

10 **(a)** $Cl_{2(g)} + 2Br^-_{(aq)} \longrightarrow 2Cl^-_{(aq)} + Br_{2(aq)}$ Equation (1); state symbols (1)
(b) Chlorine molecules gain electrons more readily (1) than bromine molecules. Therefore, chlorine is a stronger oxidising agent than bromine (1). Remember: oxidising agents are, themselves, reduced (gain electrons) during reactions.

ENTHALPY CHANGES (page 64)

1

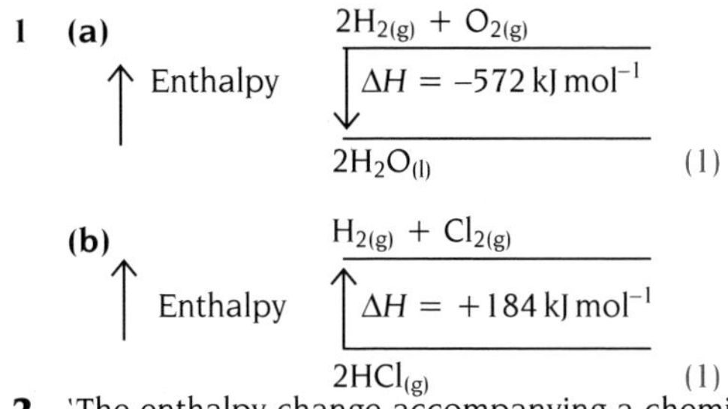

2 'The enthalpy change accompanying a chemical reaction (1) is independent of the pathway between the initial and final states' (1). Hess's law is an application of the law of conservation of energy: energy can neither be created nor destroyed.

3 **(a)** $\Delta H_f^{\ominus}$ is the enthalpy change when one mole of a compound (1) is formed, under standard conditions (1), from its elements in their standard sates (1).
(b) The calculation may be solved either by using either a Hess's law cycle or recalling a general equation. Either cycle:

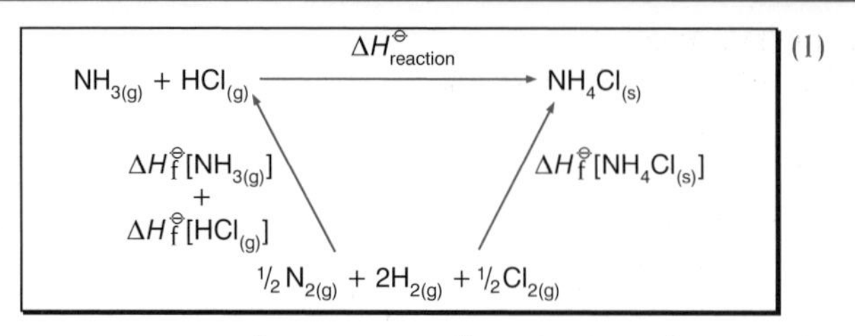

(1)

$\Delta H^{\ominus}_{reaction} = -\Delta H^{\ominus}_{f}\ [NH_{3(g)}] - \Delta H^{\ominus}_{f}\ [HCl_{(g)}] + \Delta H^{\ominus}_{f}\ [NH_4Cl_{(s)}]$
$= -(-46) - (-92) + (-314)$
$= -176$ kJ mol^{-1} (1)

Or application of equation

$\Delta H^{\ominus}_{reaction} = \Sigma\Delta H^{\ominus}_{f}$ of products $-\Sigma\Delta H^{\ominus}_{f}$ of reactants (1)
$= \Delta H^{\ominus}_{f}[NH_4Cl_{(s)}] - \{\Delta H^{\ominus}_{f}[NH_{3(g)}] + \Delta H^{\ominus}_{f}[HCl_{(g)}]\}$
$= (-314) - (-46 + -92)$
$= (-314) - (-138)$
$= -176$ kJ mol^{-1} (1)

4 **(a)** $\Delta H^{\ominus}_{c}$ is the enthalpy change when one mole of a substance (1) is completely burned in oxygen (1), under standard conditions (1).

(b) (1)

$\Delta H^{\ominus}_{f}[C_5H_{12(l)}]$
$5C_{(s,\ graphite)} + 6H_{2(g)}$ → $C_5H_{12(l)}$
$5 \times \Delta H^{\ominus}_{c}[C_{(s,\ graphite)}]$ + $6 \times \Delta H^{\ominus}_{c}[H_{2(g)}]$ $+ 8O_{2(g)}$
$+ 8O_{2(g)}$ $\Delta H^{\ominus}_{c}[C_5H_{12(l)}]$
$5CO_{2(g)} + 6H_2O_{(l)}$

From the cycle it follows that:

$\Delta H^{\ominus}_{f}\ [C_5H_{12(l)}] = 5 \times \Delta H^{\ominus}_{c}\ [C_{(s,\ graphite)}] + 6 \times \Delta H^{\ominus}_{c}\ [H_{2(g)}] - \Delta H^{\ominus}_{c}\ [C_5H_{12(l)}]$ (1)
$= (5 \times -395) + (6 \times -286) - (-3520)$
$= (-1975) + (-1716) + (3520)$
$= -171$ kJ mol^{-1} (1)

Remember to multiply $\Delta H^{\ominus}_{c}\ [C_{(s)\ graphite}]$ by five and $\Delta H^{\ominus}_{c}\ [H_{2(g)}]$ by six.

5 (1)

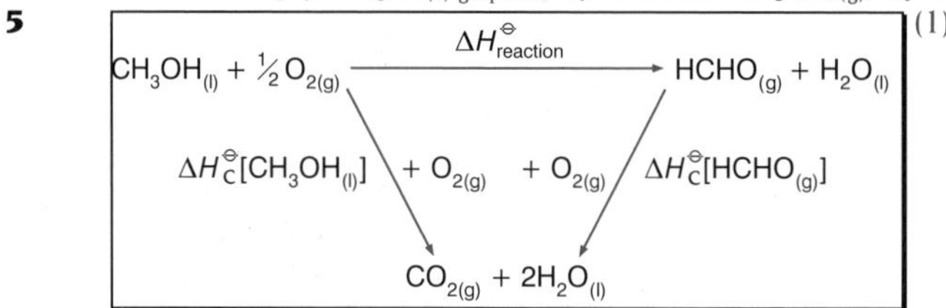

From the cycle it follows that:
$\Delta H^{\ominus}_{reaction} = \Delta H^{\ominus}_{c}[CH_3OH_{(l)}] - \Delta H^{\ominus}_{c}[HCHO_{(g)}]$

$= (-725) - (-561)$
$= -164 \text{ kJ mol}^{-1}$ (1)

Always check signs of enthalpy values are reversed when moving round the cycle in the opposite direction to an arrow.

6 (1)

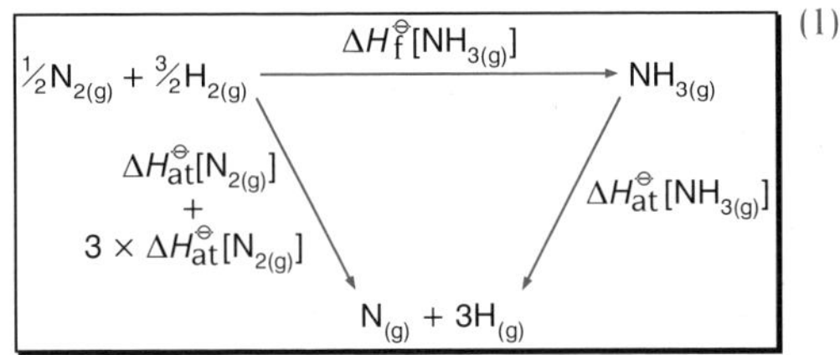

$\Delta H^{\ominus}_{at}[NH_{3(g)}] = -\Delta H^{\ominus}_{f}[NH_{3(g)}] + \Delta H^{\ominus}_{at}[N_{2(g)}] + 3 \times \Delta H^{\ominus}_{at}[H_{2(g)}]$

$= -(-46.2) + (+473) + (3 \times +218)$
$= (+46.2) + (473) + (654)$

$= +1173.2 \text{ kJ mol}^{-1}$ (1)

Each mole of $NH_{3(g)}$ contains three moles of (N–H) bonds. Therefore $\Delta H^{\ominus}_{at}[NH_{3(g)}] = 3 \times \overline{E}$(N–H) where $\overline{E}$(N–H) is the average bond enthalpy for the (N–H) bond. Therefore

$3\,\overline{E}(N\text{–}H) = +1173.2$

$\overline{E}(N\text{–}H) = \dfrac{+1173.2}{3}$

$= +391 \text{ kJ mol}^{-1}$ (3 sig. fig.) (1)

Remember $\Delta H^{\ominus}_{at}$ for an element refers to per mole of atoms produced, e.g. $\Delta H^{\ominus}_{at}[N_{2(g)}]$ is the enthalpy required for $\frac{1}{2}N_{2(g)} \rightarrow N_{(g)}$.

7 Re-write equation using structural formulae: (1)

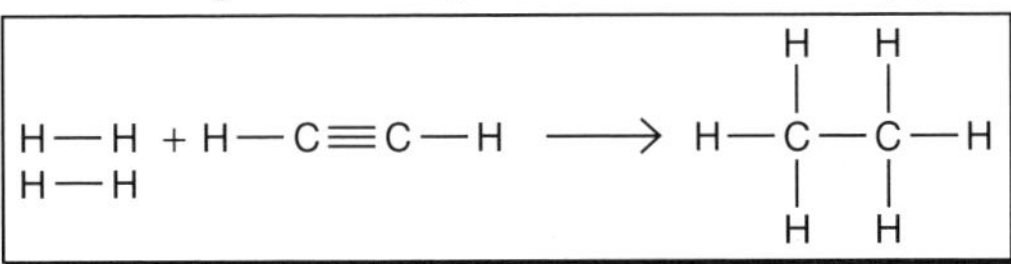

Bonds broken		Bonds made	
$2 \times \overline{E}$ (H–H)	$= +872 \text{ kJ mol}^{-1}$	$-\overline{E}$ (C–C)	$= -346 \text{ kJ mol}^{-1}$
$\overline{E}$ (C≡C)	$= +835 \text{ kJ mol}^{-1}$	$4 \times -\overline{E}$ (C–H)	$= -1652 \text{ kJ mol}^{-1}$
Total	$= +1707 \text{ kJ mol}^{-1}$	Total	$= -1998 \text{ kJ mol}^{-1}$ (1).

Add the values for bond breaking and bond making for an approximate $\Delta H^{\ominus}_{reaction}$

Check yourself answers

Enthalpy Changes (page 65)

$\Delta H^{\ominus}_{reaction} = +1707 + (-1998)$
$= -291$ kJ mol^{-1} (1)

Remember to add up the values for bond breaking and bond making, having used the correct enthalpy signs.

Reaction Rates (page 74)

1 Any four of: Concentration of a solution (1). Pressure, if the reactants are gases (1). Temperature (1). Catalyst (1). Surface area of any solids (1). Light, if the reactant is photochemical (1).

2 A reaction occurs due to collision of particles (1). The energy of the collision must exceed the activation energy for the reaction (1). The collision must occur with the correct orientation (or geometry) (1).

3 Activation energy is the minimum energy (1) that the reacting particles must possess before they can collide successfully (1).

4 On increasing concentration, there are more reactant particles per unit volume (1) do not simply put 'more reactant particles'; an increase in the frequency of the collisions (1); therefore the greater the probability of a collision with energy more than the activation energy (1) also known as a 'successful collision'.

5 Curve at lower temperature (1). Curve at higher temperature (1). This curve must be shifted to the right and have a lower peak. E_A indicated beyond both maxima (1). Shaded area beyond E_A larger for high temperature curve (1). The curves show that, at the higher temperature, the number of molecules with energy greater than or equal to E_A increases.

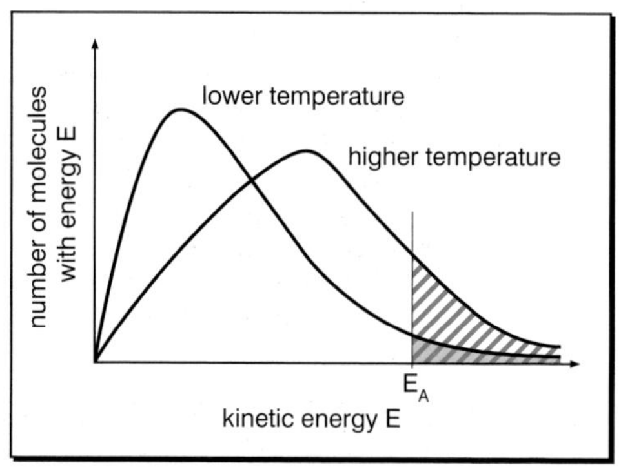

6 A catalyst is a substance that increases the rate of a reaction (1) but remains chemically unchanged at the end (1).

7 A catalyst provides an alternative route (1) with a lower activation energy (1). It is incorrect to state that a catalyst 'lowers the activation energy'.

Check yourself answers

CHEMICAL EQUILIBRIA (page 81)

1 'Dynamic': the rate of the forward reaction (A + B ⟶ C + D) (1) equals the rate of the backward reaction (C + D ⟶ A + B) (1). 'Equilibrium': the concentrations of A, B, C and D remain constant (1). Remember: do not write that the concentrations of A, B, C and D are 'equal'.

2 Concentration (1), pressure (1) and temperature (1) may, if changed, alter the position of a chemical equilibrium. These factors often, but not always, have an effect on the position of equilibrium.

3 A catalyst increases the rate of reaction as it offers an alternative reaction pathway (1) of lower activation energy (1). The position of equilibrium is not affected as the rates of both the forward and backward reactions are increased (1) to the same extent (1).

4 'If a system in equilibrium (1) is subjected to a change which disturbs the equilibrium (1), the system responds in such a way as to counteract the effect of the change (1).'

5 **(a)** Position of equilibrium shifts to the right (1). The system shifts to the side of fewer moles of gas molecules (1) (two moles on the right as opposed to three moles on the left), thereby reducing the total pressure.

(b) No effect (1) on the position of equilibrium. There are equal numbers of moles of gas molecules (1) on the left- and right-hand side.

6 **(a)** Position of equilibrium shifts to the right (1). The equilibrium moves to the exothermic direction (1). The system responds so as to produce heat energy, thereby counteracting the effect of the change.

(b) No effect (1). Because $\Delta H_{reaction}$ is zero (1) there is no exothermic or endothermic direction in this system.

7 **(a)** Equilibrium shifts to the right (1). The concentration of hydrogen ions is decreased (1) when hydroxide ions are added.
$H^+_{(aq)} + OH^-_{(aq)} \longrightarrow H_2O_{(l)}$ Equilibrium shifts to produce more $H^+_{(aq)}$ ions.

(b) Equilibrium shifts to the left (1). The concentration of ethanoic acid CH_3COOH is reduced (1) as it reacts with alkaline sodium hydroxide.
In questions 5, 6, and 7, answers such as 'by Le Chatelier's principle' are insufficient as explanations.

ORGANIC CHEMISTRY (page 94)

1 In a homologous series, the compounds have the same general formula **or** successive members of the series differ by a $-CH_2-$ unit (1). The compounds undergo similar chemical reactions **or** contain the same functional group (1).

2 **(a)** Structural isomers have the same molecular formula (1) but different structural formula (1).

(b) Geometrical isomers have the same molecular and structural formula (1) but the arrangement of their atoms in space is different (1). cis/trans isomerism is a type of geometrical isomerism encountered at AS level.

3 $CH_3CH_2CH_2CH_2CH_3$ pentane (1)
$CH_3CH_2CH(CH_3)CH_3$ 2-methylbutane (1)
$CH_3C(CH_3)_2CH_3$ 2,2-dimethylpropane (1)
Both name and structural formula required in each case.

4 **(a)** Electron pair acceptor (1)
(b) Electron pair donor (1)
(c) A species with a single unpaired electron (1)
(d) A covalent bond breaks, with one electron of the pair going to each atom (1)
(e) A covalent bond breaks, with both electrons going to one atom (1)

5 $CH_3CH_2CH_2CH_3 + Cl_2 \longrightarrow CH_3CH_2CH_2CH_2Cl + HCl$ (1)
Conditions: u.v. light (1) Not just 'light'. **Any** H atom on butane could have been substituted.

6 $CH_3{-}CH{=}CH_2 + Br_2 \longrightarrow CH_3{-}CH(Br)CH_2Br$ (1)
Conditions: bromine in an inert solvent at r.t.p. (1)

7 $CH_3CH(Br)CH(CH_3)CH_3 + KOH \longrightarrow CH_3CH(OH)CH(CH_3)CH_3 + KBr$ (1)
Conditions: heat under reflux (1) with aqueous potassium hydroxide (1)

8 $CH_3CH(Br)CH(CH_3)CH_3 + KOH \longrightarrow CH_2{=}CHCH(CH_3)CH_3 + KBr + H_2O$ (1)
and $CH_3CH(Br)CH(CH_3)CH_3 + KOH \longrightarrow CH_3CH{=}C(CH_3)_2 + KBr + H_2O$ (1)
Note how elimination of HBr is favoured when the solvent for the KOH is ethanol instead of water.

9 **(a)** (Homolytic) free radical (1) substitution (1)
(b) (Heterolytic) electrophilic (1) addition (1)
(c) Nucleophilic (1) substitution (1)
(d) Elimination (1)

10 $CH_3CH_2CH_2CH_2OH$ (1)
$(CH_3)_3COH$ (1)
$CH_3CH(CH_3)CH_2OH$ (1)
$CH_3CH_2CH(CH_3)OH$ (1)

11 $CH_3CH_2CH_2CH_2OH$ (1) $\longrightarrow CH_3CH_2CH_2CO_2H$ (1)
$CH_3CH(CH_3)CH_2OH$ (1) $\longrightarrow CH_3CH(CH_3)CO_2H$ (1)

$CH_3CH_2CH(CH_3)OH$ (1) $\longrightarrow CH_3CH_2C(=O)CH_3$ (1)

Primary alcohols are oxidised to carboxylic acids, whereas secondary alcoholsare oxidised to ketones.
$(CH_3)_3COH$ not oxidised (tertiary alcohol)

Check yourself answers

INDUSTRIAL CHEMISTRY (page 104)

1 The fractions have different boiling temperatures (1).

2 **(a)** $C_{19}H_{40}$ (1) Use the general formula for an alkane C_nH_{2n+2}.

(b) $C_{16}H_{34} \longrightarrow 2\ C_2H_4 + C_3H_6$ (1) $+\ C_9H_{20}$ (1). Ensure that total number of C atoms adds up to 16, and H atoms to 34, on the right-hand side.

(c) During cracking larger hydrocarbon molecules ($C_{16}H_{34}$) are broken down (1) into smaller alkane molecules (C_9H_{20}) plus alkene molecules (C_2H_4 and C_3H_6) (1). Larger alkane molecules have been broken down into smaller, more useful ones, some of which have carbon–carbon double bonds.

3 **(a)** $C_6H_{12}O_{6(aq)} \longrightarrow 2C_2H_5OH_{(aq)} + 2CO_{2(g)}$ (1)

(b) $C_2H_{4(g)} + H_2O_{(g)} \longrightarrow C_2H_5OH_{(g)}$ (1)

4 **(a)** Free radical (1) addition (1). Classification of organic reactions is required at AS level.

(b) 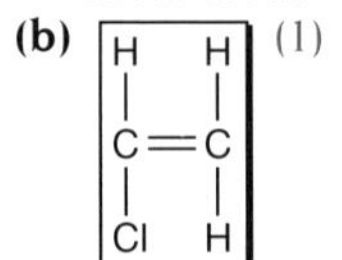 (1)

(c) 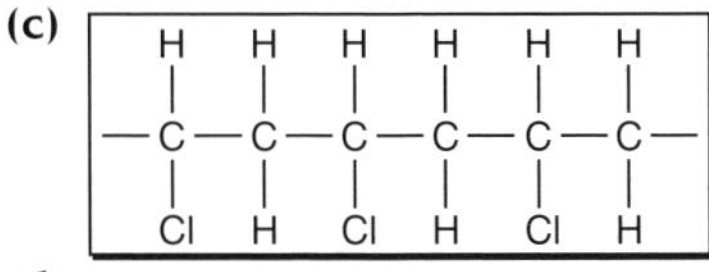 (1) Carbon–carbon double bond broken on polymer formation.

(d) 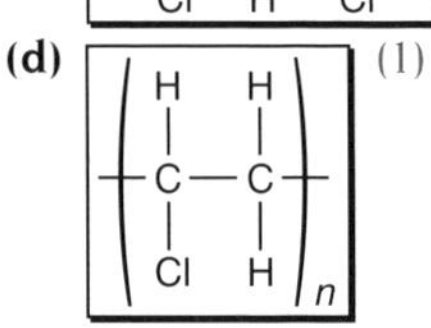 (1)

5 The very high (C–F) bond energy makes PTFE chemically inert, therefore ideal for non-stick ovenware (1). Strong (C–Cl) bonds give PVC a useful life, but creates problems for its disposal (1).

Periodic Table (Main Groups)

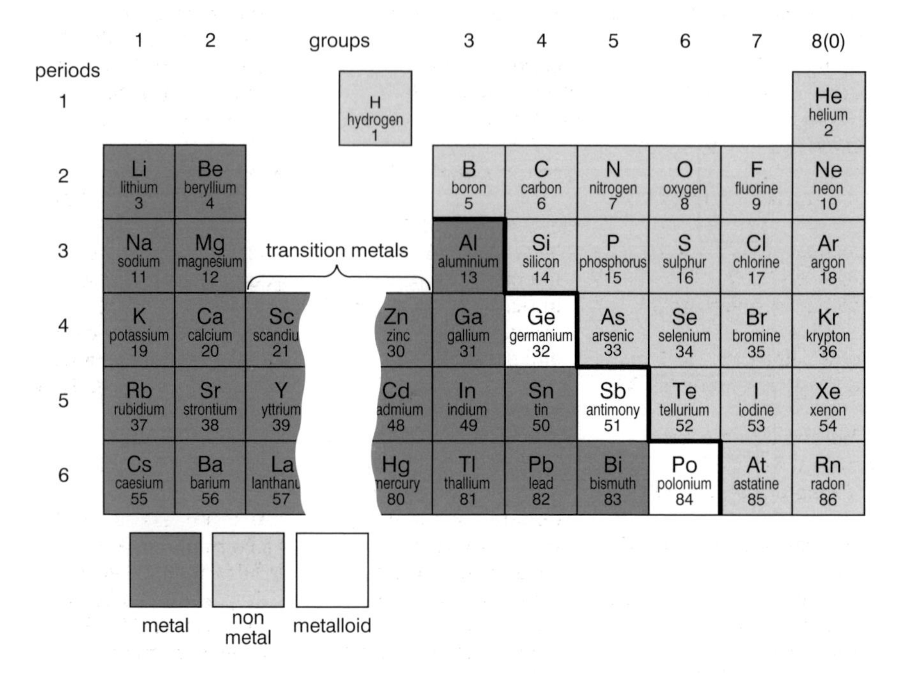

INDEX